THE
NOBEL
CHRONICLES

THE NOBEL CHRONICLES

A Handbook of Nobel Prizes in Physiology or Medicine, 1901-2024

(SECOND EDITION)

TONSE RAJU

AUCTOREM
HOUSE

Auctorem House
276 5th Ave, Ste 704-2591
New York, NY 10001
www.auctoremhouse.com
Phone: 1 888-332-7718

Published by Auctorem House: 09/29/2025

ISBN: 978-1-965687-98-7(sc)
ISBN: 978-1-965687-99-4(e)

Library of Congress Control Number: 2025920931

The Front Cover

The Spirit of Medicine Warding Off the Dragon of Disease, c. 1935, by Olga Chassaing (1897-1944) and Édouard Chassaing (1895-1974). Cast stone, 4' 7" X 5' 6". University of Illinois College of Medicine, Chicago, Illinois.

In this allegorical sculpture, the *Spirit of Medicine* carries *Infant Humanity* on her shoulder protecting it. She is sitting on the *Dragon of Disease* and warding it off by pressing her hand on its head. Her power and determination are gracefully reflected in her body language. The Chassaing couple combined Greek and pre-Columbian motifs in this beautiful artwork and dedicated it to the University of Illinois Medical Center in Chicago in 1935. Originally from France, the Chassaing couple lived in Chicago all their lives. They created this and other sculptures from grants from President Franklin D. Roosevelt's New Deal Arts Programs, that included Work Relief Program, and Federal Art Project administered by the Work Progress Administration.

This statue has been relocated multiple times. At present it is sitting on a grassy patch of land in the courtyard behind the Neuropsychiatry Institute on Wolcott Street. The university is innovating this site, after which it intends to install the *Spirit of Medicine* statue in this courtyard.

I thank Mr. Joseph Greenia, Department of Pediatrics, University of Illinois, Chicago, for photographing the statue used for this second edition cover page.

References:

1. Scheinman, M: *A Guide to the Art at the University of Illinois,* Chicago, University of Illinois Press, 1995.

2. Living New Deal: UI Medical Center, College of Medicine: Olga Chassaing Sculpture - Chicago IL - Living New Deal

Contents

To

Manu, Sharat, and Vidya
for listening to these and other
medical history tales from me
for so many years,
most often willingly,

And

To

Ananda, Sanjay, Sonya, and Kavi
hoping someday, you will read these stories
and be inspired from them

PREFACE TO THE FIRST EDITION

The Nobel Prizes are the oldest and the most prestigious of awards bestowed upon individuals for benefiting mankind at large. Perhaps, reading about the prize-winning works will be an interesting way to review the history of that subject. This handbook is offered as an introduction to some of the most important advances in medicine in the 20th century. Here I present brief notes on the prize-winning research in Physiology or Medicine from 1901-2000, along with biographical outlines for each respective laureate. The book is also a token of my personal tribute to the Nobel Foundation's first centennial celebrations in December 2001.

It was the Nobel Prizes that introduced me to the study of medical history. When I was a student, our professor once asked during ward rounds: "Which Nobel laureate discovered streptomycin?" No one in our class knew the answer—worse still, no one could name a *single* Nobel laureate of that decade. Our professor charged us that we must each in turn bring a brief presentation on any topic in medical history. Thus, we began wiping the dust off the medical history books in our library! At first, we felt the chore to be burdensome. Soon we realized that reading medical history was not only informative, but also pure fun. We learned (belatedly) that it was Selman Waksman, the winner of the 1952 Nobel Prize, who had discovered streptomycin.

When it comes to the study of medical history, today's medical students are no better than I was in my college days. Even those who profess to 'love' the subject say that they have no time for it now, and that after retiring they will study medical history. Like gardening (they seem to say), medical history is to be pursued as a hobby when there is little else to do.

But knowledge of medical history is considered most valuable *during* the active periods of one's medical practice. Yet, unfortunately, very few medical schools have medical history and ethics as required courses in their curriculum. Besides learning about the tedious nature of scientific research, the reader of science history will learn that advances do not coincide with the "6-O-clock news." Years of work, often frustrating, precedes many discoveries. Then there is the "chance factor," which might detour, delay, or hasten a particular invention or discovery. Even the greatest of scientists manage to bridge but a narrow gap in our collective knowledge. The study of medical history also shows us why some ideas that appear simple and obvious today took so many years to evolve. Similarly, we learn why bizarre and untenable notions and practices prevailed for centuries. Adding such insights *during* the years of one's practice should be an essential component of one's growth.

This book began as a minor project in 1997, when I assigned one Mr. Ranjan Ginde, then a junior in college, the task of compiling some Nobel statistics. Having read many fascinating biographies and scientific writings in preparation for a presentation on this topic, I made an ambitious proposal to *The Lancet* that if they would publish them, I could write brief essays on each of the Nobel Prize-winning works. This way, I suggested that *before* the century ended, we could cover all the 20th century's medicine prizes. *The Lancet* agreed and *The Nobel Chronicles* series was published between July 1998 and August 2000.

This book is a collection of *The Lancet* series with some modifications. There are 91 chapters dealing with as many awards for Physiology or Medicine between 1901 through 2000, along with biographies and photographs of 172 respective laureates. Each chapter begins with a list of names of laureates in *all* the categories, and their respective citations. This is followed by a biographical note on the laureate and a description of the prize-winning work. Although I have stressed their impact on contemporary medical practice and reflected upon the long-term significance of the prize-winning works, I have avoided lengthy and detailed treatment of these issues.

No biography can be narrated or discovery explained in 250 words—the space *The Lancet* allotted me. This book therefore remains a "handbook" in its strictest sense. Despite its brevity, however, the book may serve as an introduction to some of the most important medical advances of the 20th century. It may also serve as an authentic source of reference.

Without the photos of the laureates, this book would have been lifeless. I owe my gratitude to Ms. Kristina Fallenius of the Nobel Foundation and Ms. Catherine Draycott of the Wellcome Institute Library in London for providing the photos for my series in *The Lancet*, as well as allowing me to use most of them in this book.

Neither *The Nobel Chronicles* series nor this book would have been possible without the support of *The Lancet* organization and the journal's Editor in Chief, Dr. Richard Horton. I thank them both, particularly for letting me use the text from my series in this book. *The Lancet* editorial staff was also very helpful. I must especially thank Ms. Pia Pini and Ms. Joanna Palmer for their enthusiastic support during the writing of the series. I gratefully acknowledge Mr. Patrick Drazen, Dr. Tanzeema Hossain, Professor Louis Magner, Dr. Kristine McCulloch, and Mr. Sharat Raju for correcting the manuscript at various phases of its evolution. I thank Ms. Gail Bolton, Ms. Ann Dentzler, and Mr. Joseph Greenia for secretarial assistance. Mr. Michael Kirda provided the photograph of the statue, *The Spirit of Medicine*, used for the front cover—I sincerely appreciate his help.

I was fortunate to be able to use the Library of Health Sciences at the University of Illinois in Chicago, arguably one of the best of its kind in USA. Besides consulting many primary and secondary references, monographs, and articles from our library, I 'visited' the Nobel Foundation's website frequently. I have cited some of these at the end of each chapter and in the Appendix. While I remain grateful to all my sources, I alone remain responsible for all the errors. As the ever so vigilant readers of *The Lancet* did sometimes, the readers of this book too may write to me about any errors they may find I can correct those in future editions.

To paraphrase the incomparable sage H. G. Wells, my 'disqualification' for writing this book is evidently manifest. Yet, I will offer no apologies, for such work had to be done.

PREFACE TO THE SECOND EDITION

In this expanded second edition, I have covered the Nobel Prizes in Physiology and Medicine from 2001 through 2024. As before, my approach was to provide a brief description of the award-winning research, its impact on health and well-being, and to add brief biographical details of each of the Nobel laureates for the additional years added in the second edition.

The second edition has several differences from the previous edition.

My source: The sources of my text for the materials from 1901 through 2000 were the 91 brief articles I published in *The Lancet* between July 4th ,1998 through July 29th, 2000 (volumes 352—356, issues 9121 to 9227) under the title *Nobel Chronicles*. I thank *The Lancet* editorial staff for their permission to reprint them in this edition.

In the second edition, I have added the stories of medicine Noble Prizes from 2001 through 2024

The major source of my text for the science of award-winning topics was the pages of Nobel Foundation. I consulted several science journals in which the primary articles of Nobel Prize winning works were published through the National Library of Medicine webpages. I obtained biographical information of the prize-winning laureates from their Nobel Lectures or biographical information in the Nobel Foundation website.

Photographs: I have used the photographs printed in the first edition, which were also printed along with my articles in *The Lancet* (see above); the Lancet editors had kindly given permission for their use, acknowledged in the preface to the first editions. In this edition, in some cases, I have revised them, and added the Nobel Laureates from 2001 through 2024, choosing from Wikipedia. <u>List of Nobel laureates in Physiology or Medicine - Wikipedia.</u>

Appendix: I deleted all appendix materials from the first edition, because most of those are now easily available on the internet, including an analysis of the prize-winning topics, nationality of the awardees, and the bibliographies of the topics that earned Nobel Prizes.

Writing and publication credits: I did not use AI for writing *any part of the text in this book*! I did use the commercial software for editing and spellchecking.

I wish to convey my thanks to Mr. Joseph Greenia for the picture used in the cover page. I also wish to thank Ms. Cait C. Ryan, for the author's photograph she provided used in the back cover page.

I thank Auctorem House, New York, NY for publishing this volume and for their promotional efforts.

Tonse N. K. Raju, MD
Gaithersburg, Maryland, USA

INTRODUCTION

Alfred Nobel––The Man and his Dream

Since 1901, the Swedish Nobel Foundation has been awarding prizes for achievements in physics, chemistry, physiology or medicine, literature, and peace. A prize for the economic sciences was added in 1969, bringing the awards total to six. The economic prize, called the *Bank of Sweden Prize in Economic Sciences in Memory of Alfred Nobel*, is not strictly the 'original' Nobel—nevertheless, no one deems it lesser than the rest of the Nobel Prizes.

In their early days, the Nobel Prizes attracted scant attention, but today there is no recognition more prestigious and honorable than the Nobel Prize. The awardees quickly become international celebrities and receive well-deserved accolades. Their prize-winning works too are given high-profile scrutiny and an eventual place in history.

When the first set of Nobel Prizes were awarded in 1901, the human blood groups were unknown, and few if any, knew the meaning of such words as vitamins, genetics, and synapses. There were no departments of radiology or anesthesia in any hospitals, and no medical student had to study immunology, biochemistry or pharmacology as we know them today. Such common drugs as sulfa, penicillin and antihistamines were unheard of, and such tests and procedures as ECG, CT scan, and kidney transplantation were decades away from entering into clinical practice.

Thus, several advances in the 20th century indeed "conferred the greatest benefit on mankind" and many of those were honored through later Nobel Prizes. For instance: the discovery of the tuberculosis bacterium and the malaria parasite; work on poliomyelitis, typhus, and allergies; development of sulfa, penicillin, and streptomycin; the discovery of vitamins and treatment of their deficiencies; the synthesis of antihistamines, beta blockers, cortisone, and anti-cancer drugs; the invention of the electrocardiogram, CT scan, and MRI; the deciphering of the structure of the DNA molecule, and other lesser known (to the general public), but equally important discoveries in cell energy metabolism, immunology, molecular biology and physiology—all these that have become inseparable parts of today's medical practice have been awarded the Nobel Prize.

ALFRED NOBEL IN 1896

© The Nobel Foundation

Despite his fame and fortune, Alfred Nobel's life was filled with ironies. He did not formally attend a high school, let alone a college—but he was a scholar, polyglot, and an inventor. He was not greedy, yet he had a keen sense for profit. He became wealthy—very wealthy but gave away all his money. The explosives from his factories were the principal weapons of destructions and war in Europe. Yet, those weapons of destruction also aided some of the greatest constructions known to man. The Panama and Suez Canals, great railway systems worldwide, thousands of dams, irrigation projects, and urban constructions could never have taken place without the power of dynamite and other explosives. Because he manufactured weapons of war, some nations held Nobel in bitter contempt, while others, for the same reason, hailed him a genius.

Alfred Nobel was born in Stockholm in 1833. It seems that one of his ancestors chose *Nobelis* as the family name to honor the first male to receive a college degree. Alfred's father, Immanuel, an inventor and an industrialist, faced financial troubles in Sweden, because of which he moved his business to St. Petersburg, Russia when Alfred was seven. In Russia, Immanuel Nobel's military hardware and weaponry factories thrived, and his connections with the Russian military paid off.

Alfred's education in Russia was through private tutoring. In addition, his father taught him practical chemistry and business management. As a teenager, Alfred Nobel also traveled widely in Europe and in the United States, studying chemistry and learning about peoples of different nations. He became fluent in dozens of languages and perhaps began formulating the seeds of 'internationalism.'

In 1852, Alfred rejoined his father in St. Petersburg and started studying the explosive properties of nitroglycerin, a chemical that had been invented five years earlier. The Nobel family returned to Sweden in 1862 and continued the business of military hardware; they built factories in which they did research on explosives as well as manufacturing weaponry. It was here that they invented a substance that changed their fortune— they patented and called it "the dynamite."

Dynamite it was! As its worldwide sales soared, colossal wealth followed the Nobels. Alfred Nobel invented other chemicals and explosives and developed simpler methods for making synthetic rubber, leather, silk, and semi-precious stones. He held over 355 patents in different countries. He built some 100 manufacturing plants in Europe and the United States, and purchased large shares in the petrol drilling lands in the Baku region of Russia, earning the nickname the "Russian Rockefeller."

In 1870, Nobel made Paris his business headquarters but moved out of there to Italy in 1891. In France, serious ill-feelings existed about his business. Meanwhile, Nobel had maintained other operations in Sweden. He had bought the Bofors Arms Works Company and built a summer manor in Karlskoga. This little house was to play a significant role in the destiny of the Nobel fortune.

Alfred Nobel spent the last years of his life in his mansion in San Remo, Italy. His health deteriorated in the 1880s. He had constant chest pains, diagnosed by doctors as angina. It is ironic that nitroglycerin, the chemical in the dynamite that blasted away mountains and enemy bridges, was given to him to relieve his chest pains—an irony that had not escaped Mr. Nobel. Recently, it has been theorized that Nobel's symptoms were from chronic nitroglycerin toxicity from inhaling the chemical in his factories and being in contact with it.

Because of worsening health, Alfred wrote his first will in 1893. He wished that his main wealth was to be bequeathed to his heirs and that a small fraction awarded to scientists for noteworthy achievements. But, in 1895 he altered the original will, and literally scribbled this final will.

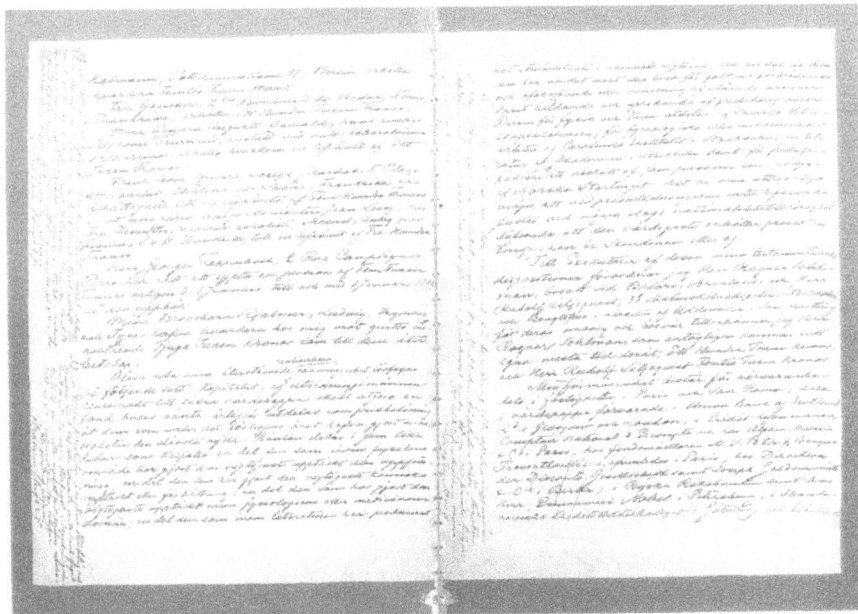

© The Nobel Foundation

The will stipulated that the *entire* Nobel holdings be liquidated and invested. From the interest, annual prizes were to be given to those who in the preceding year, "shall have conferred the greatest benefit on mankind." It identified physics, chemistry, physiology or medicine, literature and peace as the topics to be honored. The will forbade consideration whatsoever to the nationality of the nominee, especially to those of the Scandinavian descent—what mattered was the person's achievement, not the country of one's birth.

Excerpt from Alfred Nobel's Will

The whole of my remaining realizable estate shall be dealt with in the following way:

The capital shall be invested by my executors in safe securities and shall constitute a fund, the interest on which shall be annually distributed in the form of prizes to those who, during the preceding year, shall have conferred the greatest benefit on mankind. The said interest shall be divided into five equal parts, which shall be apportioned as follows: one part to the person who shall have made the most important discovery or invention within the field of physics; one part to the person who shall have made the most important chemical discovery or improvement; one part to the person who shall have made the most important discovery within the domain of physiology or medicine; one part to the person who shall have produced in the field of literature the most outstanding work of an idealistic tendency; and one part to the person who shall have done the most or the best work for the fraternity among nations, for the abolition or reduction of standing armies and for the holding and promotion of peace congresses.

The prizes for physics and chemistry shall be awarded by the Swedish Academy of Sciences; that for physiology or medical works by the Caroline Institute in Stockholm; that for literature by the Academy in Stockholm; and that for champions of peace by a committee of five persons to be elected by the Norwegian Storting. It is my express wish that in awarding the prizes no consideration whatever shall be given to the nationality of the candidates, so that the most worthy shall receive the prize, whether he be a Scandinavian or not.

Paris, November 27, 1895
ALFRED BERNHARD NOBEL

On December 10, 1896, Alfred Nobel died of cerebral hemorrhage at his home in Italy. Unmarried and childless, his love largely unrequited, he died in loneliness, with no kind hand to close his eyes or a gentle whisper to soothe his solitude. None of his romantic relationships had lasted long. Despite his wealth, Alfred Nobel was troubled by the feeling that so many had scorned his wealth as "sin money."

Why did Nobel bequeath all his wealth to a global cause? It is presumed that a strange twist of fate might have played its part. When his brother Ludwig Nobel died in 1888 a Paris newspaper mistakenly reported that Alfred had died, its headline proclaiming, *Le Marchand de la Morte est Mort* (The Merchant of Death is Dead). Alfred Nobel was understandably distraught. Did he feel compelled to alter his apparent infamous legacy, such that the Nobel family name be associated with peace and prosperity, not with death and destruction? One will never know.

How Wealthy was Mr. Nobel?

There is an amusing story concerning Alfred Nobel's wealth. When he lived in Paris, Nobel learned that his cook was getting married. He asked her what she wished to have as a wedding gift. It seems that she replied, perhaps jokingly, that "as much money as Mr. Nobel earned in one day." True to his words, Mr. Nobel meticulously calculated the sum and made out a check for that amount. On the interest generated from this gift alone, the cook lived a wealthy woman the rest of her life.

Realizing the Nobel Dream

The question of Alfred Nobel's nationality—or lack of it—became an immediate issue after his death. Consolidating the Nobel fortune as specified in the will was not easy. Nobel's business took him to various countries, where he had lived for long durations. Like many in those days, he had no passport. Lack of citizenship would have mattered little if Nobel had been poor—but his enormous wealth and the nature of his will required an immediate resolution to his ambiguous nationality. Even France, which once scorned him as "the man of death," claimed that Mr. Nobel was French because he had earned his wealth while living in France.

Strangely, a set of Nobel's horses settled the matter. It seems that a peculiar French custom had it defined that a man's 'permanent home' was the place where he had maintained his carriage horses. A few months before his death, Nobel had moved his beautiful horses to his summer home in Sweden—thus, he would be a Swede! Thus, France had no choice but to release all of Nobel's assets to Sweden, which was only too pleased to proclaim Nobel as *her* esteemed subject.

There were other hurdles. Alfred Nobel's relatives were unhappy that only 3% of his fortune was to be shared within the family and the rest went to persons unknown. Some members thought of contesting the will on the grounds that Mr. Nobel might have been mentally unsound when he wrote his final will.

A handful of his devoted friends and relatives eventually overcame all the obstacles. They ascertained and liquidated all of Nobel's assets from various nations—a tedious and complicated task by any measure. They cleared all debts and settled family grievances. Incredibly, within five years of Alfred Nobel's death, they fulfilled all the stipulations of the Nobel Will and established the Nobel Foundation. The first Nobel Prizes were awarded in 1901, marking the beginning of the new century.

The Prize

The Nobel Prize winner receives a diploma, a citation, a cash award, and the famous Nobel medal. The amount of prize money varies; it seems that at one time it was supposed to have been equal to 20 years of salary of a typical university professor! The Nobel medal is a thing of beauty: as all the medals, the Physiology or Medicine medal has a portrait of Alfred Nobel on one side and of the Genius of Medicine, a seated woman with an open book on her lap, on the other. She is filling a bowl of water from a spring to relieve the suffering and thirst of a girl standing next to her.

Some Nobel Statistics

There are many interesting stories about the Nobel Prize and the prize-winners. Jean-Paul Sartre declined the Literature Prize in 1964, and Le Duc Tho did the same of his Peace Prize in 1973. These are the only two known *voluntary* refusals of the Prize in Nobel history. Other refusals were not so voluntary.

In 1958, USSR's Boris Pasternak was forced to decline the Literature Prize for unspecified reasons. During World War II, Adolf Hitler forbade Germans to accept the prestigious award. In 1935, he forced the Peace Prize winner Carl von Ossietzky (a pacifist journalist who resisted Nazism) to decline the award. Germany's Richard Kuhn in 1938 and Adolf Butendandt in 1939 were forced to refuse *their* respective chemistry prizes. In 1939, Gerhard Domagk was not allowed to receive his Physiology or Medicine award he had earned for the discovery of the sulfa drug, Prontosil. These German winners eventually received their respective prizes after World War II. Although they acquired their medals and citations belatedly, they did not receive the prize money. As per the Nobel Foundation statute, all unclaimed prize money had to be incorporated into the general fund.

Even school children know that Marie Curie became the first woman scientist to win a Nobel Prize in 1903 in the Physics category for the discovery of radioactivity. But few know that her husband, Pierre, too shared the award the same year. Marie won the Nobel Prize again in 1911, this time in Chemistry, for discovering radium and polonium, and isolating radium.

Three others have received more than one Nobel Prize. Linus Pauling won the Chemistry Prize in 1954 and the Peace Prize in 1962—the only scientist since Marie Curie to win two Prizes in *different* fields. Frederick Sangerö of Great Britain won the 1958 Chemistry Nobel Prize for discovering the chemical structure of insulin and again in 1980 for Chemistry for work related to the DNA molecule. The American physicist John Bardeen won two Physics Prizes, the first in 1956 for transistor research, and the second in 1972 for the theory of superconductivity. Sangerö and Bardeen are the only scientists to have won two Nobel Prizes in the *same* category.

Besides the Curie couple, five-other husband-wife pairs have won Nobel Prizes. Irene Joliot-Curie (Marie Curie's daughter), and her husband Frederic Joliot shared the 1935 Prize in Chemistry; Gerty Cori and her husband Carl Cori shared the Medicine Prize in 1947; and Gunnar (Karl) Myrdal won the Economics Nobel Prize in 1974, and his wife Alva Myrdal won the Nobel Peace Prize in 1982. In 2014, May-Britt Moser and Edward Moser were jointly awarded the medicine prize for their discoveries of cells that constitute the brain's 'GPS system.' In 2019, Esther Duflo and Abhijit Banerjee were added to the list of prize-awarded couples when they were awarded the prize in economic sciences alongside their collaborator Michael Kremer.

The Curies are the only mother-daughter pair to win these famous awards. Similarly, Sir William Henry Bragg and Sir William Lawrence Bragg, were awarded the 1915 Nobel Prize in Physics for their work on X-ray crystallography—the only father-son pair to share a Nobel Prize in the *same* year. The 1970 medicine Nobel Prize winner Ulf von Euler's father, Hans von Euler-Chelpin, had won the 1929 Nobel Prize in Chemistry. Interestingly, von Euler-Chelpin's godfather, Svante Arrhenius, also had won the 1903 Chemistry Nobel Prize.

Renowned physicist Niels Bohr and his son, Aage Bohr, are perhaps the best-known father-son pair to receive the award. Niels Bohr received his award for physics in 1922. In 1975, Aage Bohr shared the Nobel Prize in Physics with Ben Roy Mottelson and James Rainwater «for the discovery of the connection between collective motion and particle motion in atomic nuclei and the development of the theory of the structure of the atomic nucleus based on this connection».

Another father-son duo to become Nobel Prize laureates was Sune K. Bergström and Svante Pääbo. Bergström was awarded the Nobel Prize in Physiology or Medicine in 1982 for producing pure prostaglandins – hormone-like substances used as medicines. 40 years later, in 2022, his son received the prize for his discoveries concerning the genomes of extinct hominins and human evolution.

Although many awards are 'shared' by individuals, only the

Peace Prize has been given to institutions; some notable ones are: the International Red Cross, the United Nations High Commission for Refugees, Permanent International Peace Bureau, Amnesty International, International Labor Organization, International Physicians for Prevention of Nuclear War, the United Nations Children's Fund (UNICEF), the United Nations Peace Keeping Force; and *Médicine sans Frontières* (Doctors Without Borders.)

I have presented additional data on the Nobel Prizes in Physiology or Medicine in the postscript articles.

Nobel Prize Winners in Physiology or Medicine 1901-2000

Emil Adolf von Behring	1901
Ronald Ross	1902
Neils Ryberg Finsen	1903
Ivan Petrovich Pavlov	1904
Robert Koch	1905
Camillo Golgi Santiago Ramon y Cajal	1906
Alfonse Alphonse Laveran	1907
Paul Ehrlich Élie Metchnikoff	1908
Theodor Kocher	1909
Albrecht Kossel	1910
Allvar Gullstrand	1911
Alexis Carrel	1912
Charles Richet	1913
Robert Bárány	1914
No Medicine Awards 1915-1919	
Jules Bordet	1919
August Krogh	1920
No Medicine Awards 1921	
Archibald Vivian Hill Otto Fritz Meyerhof	1922

Sir Frederick G. Banting	1923
John J. R. Macleod	
Wilhelm Einthoven	1924
No Medicine Awards 1925	
Johannes Fibiger	1926
Julius Wagner-Jauregg	1927
Charles Nicolle	1928
Christiaan Eijkman	1929
Sir Frederick Hopkins	
Karl Landsteiner	1930
Otto Heinrich Warburg	1931
Edgar Douglas Adrian	1932
Sir Charles Scott Sherrington	
Thomas Hunt Morgan	1933
George Richards Minot	1934
William Parry Murphy	
George Hoyt Whipple	
Hans Spemann	1935
Sir Henry Hallett Dale	1936
Otto Loewi	
Albert Szent-Györgyi	1937
Corneille Jean François Heymans	1938
Gerhard Domagk	1939
No Awards in Any Category: 1940-42	
Henrik Dam	1943
Edward Adelbert Doisy	
Joseph Erlanger	1944
Herbert Spencer Gasser	
Ernst Boris Chain	1945
Sir Alexander Fleming	
Baron Florey	
Hermann Joseph Muller	1946

Carl F. Cori	1947
Gerty T. Cori	
Bernardo Alberto Houssay	
Paul Hermann Müller	1948
Walter Rudolf Hess	1949
Antonio Egas Moniz	
Philip Showalter Hench	1950
Edward Calvin Kendall	
Tadeus Reichstein	
Max Theiler	1951
Selman Abraham Waksman	1952
Sir Hans Adolf Krebs	1953
Fritz Albert Lipmann	
John Franklin Enders	1954
Frederick Robbins	
Thomas Weller	
Axel Hugo Theodor Theorell	1955
André Frédéric Cournand	1956
Werner Forssmann	
Dickinson W. Richards, Jr.	
Daniel Bovet	1957
George Wells Beadle	1958
Joshua Lederberg	
Edward Lawrie Tatum	
Arthur Kornberg	1959
Severo Ochoa	
Sir Macfarlane Burnet	1960
Peter Brian Medawar	
Georg von Békésy	1961

Francis Crick	1962
James D. Watson	
Wilkins, Maurice H. F.	
Sir John Carew Eccles	1963
Alan Lloyd Hodgkin	
Andrew F. Huxley	
Konrad E. Bloch	1964
Feodor Lynen	
Francois Jacob	1965
André Lwoff	
Jacques Lucien Monod	
Charles Brenton Huggins	1966
Peyton Rous	
Ragnar Granit	1967
Haldan Keffer Hartline	
George Wald	
Robert W. Holley	1968
Har Gobind Khorana	
Marshall W. Nirenberg	
Max Delbrück	1969
Alfred Day Hershey	
Salvador Edward Luria	
Julius Axelrod	1970
Bernard Katz	
Ulf S Von Euler	
Earl Wilbur, Sutherland, Jr.	1971
Gerald M. Edelman	1972
Rodney Robert Porter	
Karl von Frisch	1973
Konrad Lorenz	
Nikolaas Tinbergen	
Albert Claude	1974
Christian De Duve	
George Emil Palade	

David Baltimore Renato Dulbecco Howard M. Temin	1975
Baruch S. Blumberg D. Carleton Gajdusek	1976
Roger Guillemin Andrew Victor Schally Rosalyn S. Yalow	1977
Werner Arber Daniel Nathans Hamilton O. Smith	1978
Allan M. Cormack Sir Godfrey Newbold Hounsfield	1979
Baruj Benacerraf Jean Dausset George D. Snell	1980
David H. Hubel Roger W. Sperry Torsten N. Wiesel	1981
Sune K.D. Bergström Bengt I. Samuelsson John R. Vane	1982
Barbara McClintock	1983
Niels K. Jerne George J. F. Köhler César Milstein	1984
Michael S. Brown Joseph L. Goldstein	1985
Stanley Cohen Rita Levi-Montalcini	1986
Susumu Tonegawa	1987

James Whyte Black	1988
Gertrude Belle Elion	
George Herbert Hitchings	
J. Michael Bishop	1989
Harold E. Varmus	
Joseph E. Murray	1990
E. Donnall Thomas	
Erwin Neher	1991
Bert Sakmann	
Edmond H Fischer	1992
Edwin Krebs	
Richard John Roberts	1993
Phillip A. Sharp	
Alfred G. Gilman	1994
Martin Rodbell	
Edward B Lewis	1995
Christiane Nüsslein-Volhard	
Eric Francis Wieschaus	
Peter Charles Doherty	1996
Rolf M. Zinkernagel	
Stanley Ben Prusiner	1997
Robert Francis Furchgott	1998
Louis Ignarro	
Ferid Murad	
Günter Blobel	1999
Arvid Carlsson	2000
Paul Greengard	
Eric Kandel	

Leland H. Hartwell	2001
Sir Tim Hunt	
Sir Paul M. Nurse	
Sydney Brenner	2002
H. Robert Horvitz	
Sir John E. Sulston	
Paul Lauterbur	2003
Sir Peter Mansfield	
Richard Axel	2004
Linda B. Buck	
Barry J. Marshall	2005
J. Robin Warren	
Andrew Z. Fire	2006
Craig C. Mello	
Mario R. Capecchi	2007
Sir Martin Evans	
Oliver Smithies	
Harald zur Hausen	2008
Françoise Barré-Sinoussi	
Luc Montagnier	
Elizabeth H. Blackburn	2009
Carol W. Greider	
Jack W. Szostak	
Sir Robert G. Edwards	2010
Bruce A. Beutler	2011
Jules A. Hoffmann	
Ralph M. Steinman	
Sir John B. Gurdon	2012
Shinya Yamanaka	
James E. Rothman	2013
Randy W. Schekman	
Thomas C. Südhof	
John O'Keefe	2014
May-Britt Moser	
Edvard I. Moser	

William C. Campbell	2015
Satoshi Ōmura	
Tu Youyou	
Yoshinori Ohsumi	2016
Jeffrey C. Hall	2017
Michael Rosbash	
Michael W. Young	
James P. Allison	2018
Tasuku Honjo	
William Kaelin Jr.	2019
Peter J. Ratcliffe	
Gregg L. Semenza	
Harvey J. Alter	2020
Michael Houghton	
Charles M. Rice	
David Julius	2021
Ardem Patapoutian	
Svante Pääbo	2022
Katalin Karikó	2023
Drew Weissman	
Victor Ambros	2024
Gary Ruvkun	

1901

Physics
WILHELM CONRAD RONTGEN *in recognition of the extraordinary services he has rendered by the discovery of the remarkable rays subsequently named after him.*

☆

Chemistry
JACOBUS HENRICUS VAN'T HOFF

in recognition of the extraordinary services he has rendered by the discovery of the laws of chemical dynamics and osmotic pressure in solutions.

☆

Physiology or Medicine
EMIL ADOL VON BEHRING

for his work on serum therapy, especially its application against diphtheria, by which he has opened a new road in the domain of medical science and thereby placed in the hands of the physician a victorious weapon against illness and deaths

☆

Literature
SULLY PRUDHOMME

in special recognition of his poetic composition, which gives evidence of lofty idealism, artistic perfection and a rare combination of the qualities of both heart and intellect.

☆

Peace
JEAN HENRI DUNANT

founder of the International Committee of the Red Cross, Geneva; Initiator of the Geneva Convention, and

☆

FREDÉDÉRIC PASSY

founder and President of the first French Peace Society

1901

EMIL ADOLF VON BEHRING (1854-1917)

© The Nobel Foundation

Emil Adolf von Behring received the first Nobel Prize in Physiology or Medicine for "his work on serum therapy, especially its application against diphtheria."

Behring was born into a family of modest means in Hansdorf, Prussia, where his father was schoolmaster. Those days, students had to join either the army or the church to receive a free education in universities. Behring was lucky to receive some financial help from a family friend, with whose encouragement Behring could join the Pépinière Military Institute of Medicine in Berlin. He graduated from there as an army doctor in 1878 and served in the military for about 7 years.

Like all infectious diseases diphtheria was a terror in the 19th century. It killed some seventy thousand children annually in Germany alone. It seems that in one small town (in Silesia) where Behring initially worked as doctor, a major epidemic of diphtheria is said to have made a deep impression on him; he determined to pursue research to conquer this disease.

In 1889 Behring joined Robert Koch's Institute of Hygiene and collaborated with Shibasaburo Kitasato—a Koch pupil who had discovered the tetanus bacillus. Behring and Kitasato showed that animals injected with tetanus bacteria produced, in their serum, chemicals that could kill tetanus bacilli; they named the chemicals "antitoxins."

Another Koch student, Friedrich Loeffler, had discovered the diphtheria bacillus in 1884. Adopting the tetanus model, Behring tried developing diphtheria antitoxin. After numerous failures a small quantity was produced, which protected susceptible animals from diphtheria. On Christmas night in 1891 Behring used it successfully on an 8-year-old diphtheria patient, the first to be saved using a specific treatment for diphtheria.

The public welcomed the discovery, but scientists were skeptical; however clinical trials soon established the value of serum therapy. In 1910, Behring developed a toxin-antitoxin (T-A) mixture to induce permanent immunity, or according to his own statement, 'to stimulate the body's internal disinfection.'

One- hundred years ago diphtheria devastated families, wiping out towns and villages. Behring's discovery was the first hope for its cure. Today, with immunization, diphtheria is rare, however it does strike the non-immune victims in countries where immunization routines are not carried out. Behring's methods were widely adopted in later microbiology research, and a new discipline, "immunology" began to take shape as the century progressed.

References

1. Behring EA (with S. Kitasato); "The Mechanism of Immunity to Diphtheria and Tetanus in Animals," in Milestones in Microbiology, 1961

2. Browning Carl H. Emil Behring and Paul Ehrlich. Nature 175, 1955.

3. Foster WD. A History of Medical Bacteriology and Immunology. London: Heinemann Medical, 1970.

4. Williams G. The Plague Killers. New York: Charles Scribner's Sons, 1969.

1902

☆

Physics

HENDRIK ANTOON LORENTZ **and** PIETER ZEEMAN

in recognition of the extraordinary service, they rendered by their research into the influence of magnetism upon radiation phenomena.

☆

Chemistry

HERMANN EMIL FISCHER

in recognition of the extraordinary services, he has rendered by his work on sugar and purine syntheses.

☆

Physiology or Medicine

SIR RONALD ROSS

for his work on malaria, by which he has shown how it enters the organism and thereby has laid the foundation for successful research on this disease and methods of combating it.

☆

Literature

CHRISTIAN MATTHIAS THEODOR MOMMSEN

the greatest living master of the art of historical writing, with special reference to his monumental work, A History of Rome.

☆

Peace

ÉLIE DUCOMMUN

Honorary Secretary of the Permanent International Peace Bureau, Berne; and

CHARLES ALBERT GOBAT

Secretary General of the Inter-Parliamentary Union, Berne. Honorary Secretary of the Permanent International Peace Bureau, Berne.

1902

SIR RONALD ROSS (1857-1932)

© The Nobel Foundation

Ronald Ross received the Nobel Prize in Physiology or Medicine "for his work on malaria." He was born in Almora, India. Instead of studying literature and the arts as he had wanted, he followed his father's wish and decided to study medicine. Because malaria was rampant worldwide, with 5 million annual deaths in India alone, Ross developed a keen interest in malaria research from very early in his career.

In 1882, Alphonse Laveran had discovered the *plasmodium* parasite, but its life cycle and modes of transmission were unknown. While working in Secunderabad, a twin city near the South Indian town of Hyderabad, Ross began to examine blood specimen from his patients with malaria and attempted to find the parasite Laveran had discovered. But Ross could not see any larvae in the blood specimen of his patients.

During a visit to England in 1894, Ross called upon Dr. Patrick Manson, the famous scientist who had discovered the larvae of filarial worms responsible for elephantiasis. Manson showed Ross, the crescent-shaped bodies of Laveran, and a variety of different forms of the malarial parasite in the stained blood specimen from patients at Charing Cross Hospital. Manson had theorized that malarial was transmitted when humans drank water with mosquito eggs along with the larval forms of the parasite. He had felt that the parasites initially entered the mosquito when they bit humans infected with the disease. As per Manson, the larval forms along with mosquito eggs were excreted into ponds, contaminating the drinking-water.

Ross initially attempted to find malarial parasites in water but failed to find any. His attempts to transmit malaria by making the volunteers to drink 'mosquito-contaminated' water also failed similarly. However, with his ordinary microscope, Ross *was* able to see the previously known flagellar forms of the parasite in mosquito stomachs. Yet, those mosquitoes, the *Aëdes* and *Culex* species, failed to transmit malaria into healthy volunteers.

In 1895, Ross was working in the Begunpet Hospital at Secunderabad. By then he had developed ingenious experimental methods for his studies. He made healthy volunteers and those with malaria remain in the same rooms that also contained mosquitoes. Upon getting bitten, the volunteers were asked to catch the mosquitoes and bring them to Ross for examination. Using a special dissection method, Ross then attempted to find various forms of the parasite in the mosquito stomachs. (Needless to say, that Ross's research methods would not have met the ethical standards these days, and it is unlikely that any Institutional Review Board would have approved them!)

In August of 1897, Ross' assistants caught ten mosquitoes of a rare form with spotted wings—the *Anopheles* species. Ross failed to see anything in the stomachs of the first eight of these bugs. On the 20th of August, when Ross was down to his final two—the 9th and the 10th mosquitoes—he discovered in their stomachs, "cells contain[ing] the pigment...[of] hemamoeba," the oocyst stage of the protozoa, which he had never seen there before. Ross then formed the brilliant conclusion that these were the parasites' fertilized gametes, a hitherto unknown stage. Two years later he proved that the parasites migrate to the salivary glands and infect animals (birds) via the saliva during a mosquito bite; he proposed a similar mechanism for malaria transmission to humans.

When Ross presented his results to the British Medical Society and published the results of his discovery, the world was caught by surprise and delight. Not only was the mysterious life cycle of the malaria parasite no longer a mystery, but also Ross' findings provided the possibility of controlling mosquitoes as a means of controlling malaria. Widespread malaria control programs soon began. Ross helped plan and supervise many such programs in England, India, and elsewhere.

Although malaria remains a global endemic, Ross's discovery was pivotal in understanding the pathogenesis of malaria. His findings helped control this devastating disease to a considerable extent through major public health control strategies. August 20 is remembered as the "Mosquito Day" in Secunderabad and elsewhere in the world.

References

1. De Kruif, Paul. Men against Death. New York: Harcourt, Brace, 1932.
2. Mégroz, Rodolphe L. Ronald Ross: Discoverer and Creator, London: Allen & Unwin, 1931
3. Ross R. (Novels): The Child of Ocean, 1889; The Deformed Transformed, 1890; The Spirit of Storm, 1896; The Revels of Orsera, 1920
4. Ross R: On some peculiar, pigmented cells found in two mosquitoes fed on malarial blood. British Medical Journal, 18th 1897.
5. Ross R: Report of the Malaria Expedition of the Liverpool School of Tropical Medicine, 1900; Malarial Fever.

1903

☆

Physics
ANTOINE HENRI BECQUEREL
in recognition of the extraordinary services, he has rendered
by his discovery of spontaneous radioactivity, and
PIERRE CURIE and MARIE CURIE (SKLODOWSKA)
in recognition of the extraordinary services, they have rendered by their joint
research on the radiation phenomena discovered by Professor Henri Becquerel

☆

Chemistry
SVANTE AUGUST ARRHENIUS
in recognition of the extraordinary services, he has rendered to the
advancement of chemistry by his electrolytic theory of dissociation.

☆

Physiology or Medicine
NIELS RYBERG FINSEN
in recognition of his contribution to the treatment of diseases,
especially lupus vulgaris, with concentrated light radiation,
whereby he has opened a new avenue for medical science.

☆

Literature
BJØRNSTJERNE MARTINUS BJØRNSON
as a tribute to his noble, magnificent and versatile poetry,
which has always been distinguished by both the freshness
of its inspiration and the rare purity of its spirit.

☆

Peace
SIR WILLIAM RANDAL CREMER
Member of the British Parliament. Secretary of
the International Arbitration League.

1903

NIELS RYBERG FINSEN (1860-1904)

© The Nobel Foundation

Niels Ryberg Finsen was awarded the Nobel Prize in Physiology or Medicine "in recognition of treatment of ...lupus vulgaris, with concentrated light rays..."

A descendent from a Viking family, Finsen was born in Thorshavn, Faeroe Islands in the North Sea between the Shetlands and Iceland. His father was the governor of the island. Young Niels studied at Thorshavn and later in Herlufsholm in Denmark. After further schooling in Reykjavík, Iceland, he joined the University of Copenhagen and graduated in medicine in 1891.

Niels Finsen suffered from debilitating Pick's Disease from childhood. Since sunlight seemed to improve his health, he was constantly preoccupied with the sun. Thus, after graduating in medicine, he explored the biological value of sunlight through research on laboratory animals and later patients. He devised "red rooms" with window curtains to block sun's ultraviolet rays and treated the skin lesions of smallpox patients. The net result, he reasoned, would be a benefit from sunlight without irritation from the full spectrum of the sun's light.

By advancing these concepts, he developed a lamp in 1895, with which he treated a friend who had lupus vulgaris. Within a few months of lamp light treatment, the skin lesions were completely cured. Finsen then assembled a torch with a carbon arc light source and lenses for focusing. With this Finsen-Ryan Lamp he treated over 800 patients with lupus vulgaris during the first year alone in his Phototherapy Institute in Copenhagen. Eighty percent of his patients were cured and Finsen's fame spread quickly.

Unfortunately, Pick's Disease took its toll. In 1903 he was too ill to travel and receive his Nobel Prize in Stockholm.

Finsen was thus a pioneer scientist exploring the value of phototherapy and the effects of sunlight spectrum on human health. His work led to the founding of the 'School of the Sun' over the Alps for tuberculosis patients. This became a major sanitarium for the outdoor treatment of tuberculosis patients; here the famous German writer Thomas Mann was treated. Mann used his experience there and the mountain backdrop in his novel *The Magic Mountain*.

Since the advent of anti-tuberculosis drugs, phototherapy has lost its therapeutic place for patients with skin TB; yet its use has been extended into many other fields in dermatology, and it remains an important adjunct in dermatological treatment today. With the Finsen stamp in 1960, the Danish Government honored this pioneer of phototherapy, 'the man who loved the sun.'

References

1. Bailey H: The Vitamin Pioneers. Emmaus, Pa: Rodale Books, 1968.
2. Broadthagen H. Stamps Commemorating Medicine "Niels Finsen" Physician, Photobiologist, Nobel Laureate. *J Dermatol Surg Oncol* 5:8, 649 1979
3. De Kruif P. Men Against Death. New York: Harcourt, Brace & World, 1932.
4. Finsen NR: Ugeskrift for Laeger, no. 7, p. 145, no. 8, p. 173, 1904.
5. Florkin M. A History of Biochemistry. Amsterdam: Elsevier, 1972.

1904

☆

Physics
LORD JOHN WILLIAM STRUTT RAYLEIGH

*for his investigations of the densities of the most important gases
and for his discovery of argon in connection with these studies.*

☆

Chemistry
SIR WILLIAM RAMSAY

*in recognition of his services in the discovery of the inert gaseous elements
in air, and his determination of their place in the periodic system.*

☆

Physiology or Medicine
IVAN PETROVICH PAVLOV

*in recognition of his work on the physiology of digestion, through which
knowledge on vital aspects of the subject has been transformed and enlarged.*

☆

Literature
FRÉDÉRIC MISTRAL

*in recognition of the fresh originality and true inspiration of
his poetic production, which faithfully reflects the natural
scenery and native spirit of his people, and, in addition, his
significant work as a Provenal Provençal philologist;* and

JOSÉ ECHEGARAY Y EIZAGUIRRE

*in recognition of the numerous and brilliant compositions
which, in an individual and original manner, have
revived the great traditions of the Spanish drama.*

☆

Peace
INSTITUTE OF INTERNATIONAL LAW

*for its efforts in establishing the law of nations as a "collective scientific action"
and for its aspiration "to serve as...the legal conscience of the civilized world."*

1904

Ivan Petrovich Pavlov (1849-1936)

© The Nobel Foundation

Ivan Petrovich Pavlov won the Nobel Prize in Physiology or Medicine "for demonstrating the role of the brain in controlling digestion and bringing about a 'profound transformation' in the theoretical knowledge in physiology."

Like his father, Ivan Pavlov should have become a priest. Discovering an interest in biology, he left the seminary. After a brilliant student career, which earned him a medical degree and a doctorate, he devoted full-time to physiology research. In 1884 he studied digestive physiology as an unpaid lecturer at the St. Petersburg Military Medical Academy, and later at the Institute of Experimental Medicine.

He developed animal models for chronic studies. He argued that in the 'ordinary' acute experimental models, there would be many errors, since the process of preparing the animal would affect its physiological state. Thus, he devised different strategies to develop 'chronic' experimental models.

To study the method of digestion over time, he created fistulae, or surgical 'windows,' in his experimental animals. These windows allowed him to collect separately the secretions from the salivary glands, stomach, pancreas, and the small intestine over several days and weeks. From these studies he concluded that secretions from the salivary gland, esophagus, duodenum, pancreas, and the liver contributed different kinds of digestive juices, and they were controlled by the brain. Thus, was born the doctrine of 'nervism,' of which Pavlov became a supreme champion. Ironically, however, much of nervism was disproved in his own time.

However, it was Pavlov's work on the physiology of reflexes that brought him fame. The method of preparing animal models for chronic study was also his enduring legacy. His research on the "psychic secretion," the Pavlovian dog, was the foundation of understanding conditioned reflex, which influenced research in physiology, general and educational psychology, psychiatry, and neurology. Many of his major conclusions in classical digestive physiology have stood the test of time.

Just as he was a tenacious and meticulous researcher, Pavlov was also a fervent anti-communist and a free thinker. The Soviet Government, however, tolerated him and his philosophies during his later years. This was in part because of Pavlov's international fame and stature as well as his 'Darwinian philosophical bent,' which explained the workings of the body and mind in 'mechanistic' terms that suited the communist philosophy. Pavlov's political views aside, his scientific legacy continues to influence the field of neurobiology and classical physiology.

References

1. Babkin BP. Pavlov: A Biography. Chicago: University of Chicago Press, 1949.
2. Cuny H. Ivan Pavlov: The Man and His Theories. Translated by Patrick Evans. London: Souvenir Press, 1962.
3. Gray JA. Ivan Pavlov. New York: Viking Press, 1980.
4. Pavlov IP. Conditioned Reflexes: An Investigation of the Physiological Activity of the Cerebral Cortex, 1927; Lectures on Conditioned Reflexes, 1928.
5. Pavlov IP. Lektsii of rabotie glavnykh pishchevaritl' nykh zhelez. 1897 (The Work of the Digestive Glands, 1902).

1905

☆

Literature
HENRYK SIENKIEWICZ

because of his outstanding merits as an epic writer.

☆

Chemistry
JOHANN FRIEDRICH WILHELM ADOLF VON BAEYER

*in recognition of his services in the advancement of organic
chemistry and the chemical industry, through his work
on organic dyes and hydroaromatic compounds.*

☆

Physics
PHILIPP EDUARD ANTON LENARD

for his work on cathode rays.

☆

Physiology or Medicine
ROBERT KOCH

for his investigations and discoveries in relation to tuberculosis.

☆

Peace
BARONESS BERTHA SOPHIE FELICIE von SUTTNER

Honorary President of the Permanent International Peace Bureau,
Berne. Author of Die Waffen Nieder (Lay Down Your Arms)
For her sincere peace activities.

1905

ROBERT KOCH (1843-1910)

© The Nobel Foundation

Robert Koch received a Nobel Prize in Physiology or Medicine "for his investigations and discoveries in regard to tuberculosis."

Son of a mining officer, Robert Koch was born in Klausthal, Germany, and wished to become an explorer. He also dreamt of winning the Iron Cross by working as a military surgeon. After a medical degree from the Göttingen he served briefly in the military during the Franco-Prussian war and later worked as district medical officer in Wollstein (now in Poland).

Koch had a busy practice that required him to make house calls on horseback. Yet, using wooden splinters, glass plates, a simple camera, and a microscope his wife gave him for his 28th birthday (so that 'he could play with it'), Koch pursued experiments in his 'home laboratory.' He became an expert in archaeology, anthropology, photography, and in the new field of bacteriology studying algae, fungi and parasites. He developed ingenious staining methods for identifying and photographing bacteria and mastered the fine art of adjusting light and focusing the microscope.

In an experiment in the 1870s, Koch placed on a slide a drop of blood from a cow dying from anthrax. Under the microscope, he saw in the slide many objects that looked like silk threads. He wondered if these 'threads' were alive and experimented by injecting them into laboratory animals. Later he identified similar objects in tissues from other cattle sick from anthrax, but not from healthy animals. He thus concluded that the 'silky creatures' were indeed anthrax-causing microbes. By 1876, he had discovered the complete life cycle of the anthrax bacillus—the first microbe to be identified with a specific disease known to man.

In 1880, Koch took up a research job at the Imperial Board of Health in Berlin. With two assistants, Koch searched for bacterial causes for other infections and managed to transmit tuberculosis to laboratory animals using lung tissues from patients dying from "galloping consumption." Initially, he could not see the bacteria of tuberculosis. But when he stained the slide using a blue dye, he found clumps of curved microbes. Koch thus became the first human to 'see' the bacteria causing tuberculosis.

When Koch presented his discovery of the tuberculosis bacterium by reading a paper in front of the Berlin Physiological Society on March 24, 1882, the audience was stunned with disbelief.

Koch established fundamental methods for identifying infectious agents and developed strict criteria for associating them with diseases. He showed that a single micro-organism can be isolated from animals suffering from anthrax and that the disease could be reproduced in experimental hosts by infection with a pure culture of this bacterium. He then concluded that the same micro-organism could be subsequently re-isolated from the experimental host. These 'rules' came to be known as *Koch's Postulates*, which remain as pillars in microbiology and infectious diseases.

Even after receiving worldwide acclaim for his bacteriology research Koch pursued laboratory research and oversaw programs for controlling the spread of epidemics. Working in India and Africa he discovered the microbes that caused cholera and carried out much of the basic research on trypanosomiasis, malaria and plague.

Although the Nobel Foundation honored Koch for discovering the tuberculosis bacterium, many of his other discoveries also deserved additional Nobel Prizes. March 24th, the day Koch announced his discovery, has been designated by the World Health Organization (WHO) as *World TB Day*. While tuberculosis is largely controlled in the industrialized West, it remains endemic in developing countries, and in such susceptible populations as those with HIV, malnutrition and immune-compromised conditions. The WHO is renewing its battle against this ancient scourge.

References

1. Baldry, Peter: The Battle Against Bacteria. Cambridge, England: Cambridge University Press, 1976.
2. Brock, Thomas D., ed. Milestones in Microbiology. Washington, D.C.: American Society for Microbiology, 1961.
3. Robert Koch. "The Etiology of Anthrax Based on the Life History of Bacillus anthracis," in Milestones in Microbiology, 1961.
4. Robert Koch: A Life in Medicine and Bacteriology. Madison, Wis.: Science Tech, 1988.

1906

☆

Physics
SIR JOSEPH JOHN THOMSON
in recognition of the great merits of his theoretical and experimental investigations on the conduction of electricity by gases.

☆

Chemistry
HENRI MOISSAN
in recognition of the great services rendered by him in his investigation and isolation of the element fluorine, and for the adoption in the service of science of the electric furnace called after him.

☆

Physiology or Medicine
CAMILLO GOLGI and SANTIAGO RAMÓN Y CAJAL
in recognition of their work on the structure of the nervous system.

☆

Literature
GIOSUÈ CARDUCCI
not only in consideration of his deep learning and critical research, but above all as a tribute to the creative energy, freshness of style, and lyrical force which characterize his poetic masterpieces.

☆

Peace
THEODORE ROOSEVELT
President of the United States of America.
drew up the 1905 peace treaty between Russia and Japan.

1906

Santiago Ramón y Cajal (1852-1934) and Camillo Golgi (1843-1926)

Santiago Ramón y Cajal   **Camillo Golgi**

© The Nobel Foundation

Camillo Golgi (right) and **Santiago Ramón y Cajal** were the recipients of the Nobel Prize in Physiology or Medicine "in recognition of their work on the structure of the nervous system."

Camillo Golgi was born in Corteno, Italy and received medical education at the University of Pavia. He first studied the lymphatics in the brain and neuroglia and later set up a laboratory in the kitchen of a small hospital in Abbriategrasso. There in 1873, he developed the silver-impregnation technique of staining brain tissues and discovered type I and type II neuronal cells; these and the cells in the cerebellar granular layer, as well as the internal reticular apparatus of neurons are named after Golgi. He was also the first to distinguish brain sarcoma from glioma.

Ramón Y Cajal was born in Petilla, a village in the Spanish Pyrenees. After a tortuous student life under tyrannical teachers, he took to studying medicine and graduated in 1873. Following several jobs, including that of army doctor in Spanish Cuba, he obtained a chair in anatomy at Valencia in 1884. It was there that Cajal made improvements on the Golgi staining techniques and systematically studied the microscopic anatomy of the brain. He and his pupils published 237 papers over the next two decades, most convincingly establishing the 'neuronal doctrine' of brain anatomy.

Golgi and Cajal disagreed on several scientific issues and their 'battles' have become legendary. However, from the perspective of neurology and its growth, their independent research findings were complementary and discoveries monumental. From Charles Sherrington to Eric Kandel, all neuroscientists owe a great intellectual debt to Golgi and Cajal.

References

1. Cajal R. The Croonian Lecture, Proceedings of the Royal Society of London, vol. 53, pp. 444-468, 1894.

2. Cajal SR. *Nuevo concepto de las histologia de los centros nerviosos.* Revista de Ciencias Médicas 18:457; 1892

3. Garrison FH. Ramón y Cajal. Bulletin of the New York Academy of Medicine 5: pp. 483-508, 1929.

4. Golgi C. All of Golgi's works published between 1868 and 1902 are reprinted in three volumes in Opera omnia, 1903.

5. McHenry LC. Garrisons's Garrison's History of Neurology. Springfield, IL.: Charles C. Thomas, 1969.

6. Riese W. A History of Neurology. New York: MD, 1959.

7. Santini M. Golgi Centennial Symposium. New York: Raven Press, 1975.

1907

☆

Physics
ALBERT ABRAHAM MICHELSON

for his optical precision instruments and the spectroscopic and metrological investigations carried out with their aid.

☆

Chemistry
EDUARD BUCHNER

for his biochemical research and his discovery of cell-free fermentation.

☆

Physiology or Medicine
CHARLES LOUIS ALPHONSE LAVERAN

in recognition of his work on the role played by protozoa in causing diseases.

☆

Literature
RUDYARD KIPLING

in consideration of the power of observation, originality of imagination, virility of ideas and remarkable talent for narration which characterize the creations of this world-famous author.

☆

Peace
ERNESTO TEODORO MONETA

President of the Lombard League of Peace and

LOUIS RENAULT

Professor International Law, Sorbonne University, Paris
for efforts at the Hague Peace Conferences in 1899 and 1907.

1907

Charles Louis ~~Alpohonse~~
Alphonse Laveran (1845-1922)

© The Nobel Foundation

Charles Laveran received the Nobel Prize in Physiology or Medicine "in recognition of his work on the role played by protozoa in causing diseases."

Laveran was born in Paris. After graduating from Strasbourg, he enlisted in the army where he served as a medical assistant and then as an ambulance officer (1878—83) during the Franco-German war. He served in a military hospital as Chief of Epidemics before being posted to Bône, Algeria, where malaria was a major cause of death. It was here that Laveran did most of his malaria research.

Laveran was struck by the peculiar black pigments in the liver and spleen of malaria victims. Large quantities of these were also seen in the white blood cells of patients, along with free-moving round particles and crescent-shaped bodies attached to the red cells. His suspicion that these particles might be parasites was confirmed on November 6, 1880, when he discovered moving flagellae on their surface. Since the organisms were too large to be bacteria, and unlike the fungi, contained flagellae, he concluded that they must be protozoa. Within a fortnight, Laveran sent a report to the Academy of Medicine in Paris of his discovery.

Initial reluctance to accept Laveran's findings that protozoa cause malaria can be appreciated in the light of the prevailing excitement due to advances in bacteriology. Most scientists then believed that bacteria were responsible for all infections. Few of them considered protozoa as possible agents causing *any* illness, and fewer still trusted Laveran's research methods. Thus, it took Laveran several frustrating years to convince the skeptics about his discovery. Eventually, however, the truth of Laveran's findings prevailed; even Louis Pasteur was convinced of it, and Manson and Ross, two famous contemporary parasitologists, among others, endorsed Laveran's proposals and helped to advance parasitology.

References

1. Laveran A: Traité des fiévres plaustres avec la description des microbes du paludisme, 1884. (Traits of Palustrial Fever with a Description of the Palustrial Microbes, 1884)
2. Laveran A: Trypanosomes et trypanosomiases, 1904 (with F. Mesnil; Trypanosomes and Trypanosomiasis, 1912).
3. Schmidt, Gerald D., and Larry S. Roberts. Foundations of Parasitology. 2d ed. St. Louis: C.V. Mosby, 1981.
4. Williams, Greer. The Plague Killers. New York: Charles Scribner's Sons, 1969.

1908

☆

Physics
GABRIEL LIPPMANN
*for his method of reproducing colors photographically
based on the phenomenon of interference.*

☆

Chemistry
LORD ERNEST RUTHERFORD
*for his investigations into the disintegration of the elements,
and the chemistry of radioactive substances.*

☆

Physiology or Medicine
ÉLIE ILYICH MECHNIKOFF and PAUL EHRLICH
in recognition of their work on immunity.

☆

Literature
RUDOLF CHRISTOPH EUCKEN
*in recognition of his earnest search for truth, his penetrating
power of thought, his wide range of vision, and the warmth and
strength in presentation with which in his numerous works he has
vindicated and developed an idealistic philosophy of life.*

☆

Peace
KLAS PONTUS ARNOLDSON
Former Member for the Swedish Parliament. Founder of
the Swedish Peace and Arbitration League, and
FREDRIK BAJER
Member of the Danish Parliament. Honorary President of
the Permanent International Peace Bureau, Berne.

1908

Paul Ehrlich (1854-1915) and
Élie Metchnikoff (1845-1916)

Paul Ehrlich   **Élie Metchnikoff**
© The Nobel Foundation

The Nobel Prize in Physiology or Medicine was awarded to **Elie Metchnikoff** (right) and **Paul Ehrlich** for "their work on the theory of immunity."

Metchnikoff was born in a village near Kharkoff, Russia. Political turmoil and personal misfortunes forced him to travel widely for education and jobs. By 1875, the germ theory of disease had been accepted and Metchnikoff conducted extensive research to explain the mechanisms of disease and immunity. Working on comparative embryology, he discovered that the cells in starfish larvae 'attacked' newly introduced foreign bodies, such as wood splinters. In the fresh-water crustacean *Daphnia*, he found white cells preventing fungal spores from invading. He named these protective white corpuscles "phagocytes," or "the devouring cells," and proposed a similar role for human white cells.

Ehrlich was born into a wealthy family and studied and worked in several German institutions. He improved bacterial staining techniques, proposed the 'side chain' theory of immunity and methods to enhance antitoxin potency, and developed Salvarsan and Neosalvarsan, the first effective remedies for syphilis. These are only but a few of the major contributions of Ehrlich. It seems he once remarked that four (German) Gs were essential for success: *Geld* (money), *Geduld* (patience), *Geschick* (cleverness), and *Glück* (luck).

Luck eluded Metchnikoff much of his life. His brother's death from cancer and his first wife's death from TB drove him to attempt suicide twice. But late in life he would be happier. He credited yogurt for his longevity and strongly advocated it. He hoped for the day when all diseases would be conquered, and death would come only from old age.

References

1. Bauer, H. Paul Ehrlich's Influence on Chemistry and Biochemistry. Annals of the New York Academy of Sciences 59, 1954.
2. Ehrlich P. Cytotoxins and Cytotoxic Immunity. Boston Medical and Surgical Journal, vol. 150, 1904.
3. Foster, WD. A History of Medical Bacteriology and Immunology. London: Heinemann Medical, 1970.
4. Metchnikoff, E. Lectures on the Comparative Pathology of Inflammation, 1893.
5. Silverstein, AM. A History of Immunology. San Diego: Academic Press, 1989.

1909

☆

Physics
GUGLIELMO MARCONI and **CARL FERDINAND BRAUN**
in recognition of their contributions to the development of wireless telegraphy.

☆

Chemistry
WILHELM OSTWALD
in recognition of his work on catalysis and for his investigations into the
fundamental principles governing chemical equilibrium and rates of reaction.

☆

Physiology or Medicine
EMIL THEODOR KOCHER
for his work on the physiology, pathology and surgery of the thyroid gland.

☆

Literature
SELMA OTTILIA LOVISA LAGERLÖF
in appreciation of the lofty idealism, vivid imagination and
spiritual perception that characterize her writings.

☆

Peace
AUGUSTE MARIE FRANÇOIS BEERNAERT
Former Prime Minister and member of the Belgian Parliament and
member of the International Court of Arbitration at the Hague; and
BARON D'ESTOURNELLES DE CONSTANT
Member of the French Parliament. Founder and President
of the French parliamentary group for international
arbitration; Founder of the Committee for the Defense of
National Interests and International Conciliation.

1909

THEODOR KOCHER (1841—1917)

Theodor Kocher was awarded the Nobel Prize in Physiology or Medicine "for his work on the physiology, pathology, and surgery of the thyroid gland."

Kocher was born in Berne, Switzerland, and graduated in 1865. During study trips through Europe, he refined his craft, working under such leading surgeons as Demme, Lücke, Billroth, and Langenbeck. He returned to Berne in 1866 and became the only surgical assistant under Lücke, on call twenty-four hours a day. Here he would work for the next four decades, developing numerous surgical innovations and instruments.

Until the advent of antiseptic surgery, such operations as the removal of enlarged thyroid gland carried very high mortality. The physiological role of the thyroid gland was also not well understood. Kocher showed that total goiter removal led to severe forms of hypothyroidism, like that of congenital myxedema (a condition of complete absence of thyroid hormones), or "cachexia thyreopriva," as he called it. He showed that sub-total thyroidectomy or administrating the "thyroid juice" alleviated some of the symptoms of post-operative myxedema.

Other Kocher contributions include, improved surgical techniques for diseases of the stomach, duodenum and gall bladder; operations for hernia, cancer, vertebral injuries and gunshot wounds; and new methods for reducing shoulder dislocations. His innovative surgical forceps, suture materials, and other surgical appliances were in vogue for a very long time.

Kocher was a deeply religious man, with a warm personality towards his students and patients. He was also one of the most skilled craftsmen in surgery. He was deeply saddened by the outbreak of World War I. He felt that all his hopes for the betterment of mankind had been shattered due to the Great War, which also isolated him from other scientists in the world as well as from close family members in the United States.

The War also took additional tolls on Kocher's health and work. He lost most of his assistants to the German Army. Yet, he decided never to retire from surgical practice. With help from his sons and a single assistant, he continued to operate and correct his students' papers until two days prior to his death at the age of 75, on July 25, 1917.

References

1. Crapo, L. Hormones: Messengers of Life. San Francisco: W.H. Freeman, 1985.
2. Degroot, LP. Reed Larsen, Samuel Refetoff, and John Stanbury. The Thyroid and Its Diseases. New York: John Wiley & Sons, 1984.
3. Kocher T. Chirurgische Operationslehre, 1892 (Textbook of Operative Surgery, 1894.
4. Merke, F, ed. History and Iconography of Endemic Goiter and Cretinism, Norwell, Mass.: Kluwer Academic Publishers, 1984.

1910

Physics
JOHANNES DIDERIK VAN DER WAALS

for his work on the equation of state for gases and liquids.

☆

Chemistry
OTTO WALLACH

in recognition of his services to organic chemistry and the chemical industry by his pioneer work in the field of salicylic compounds.

☆

Physiology or Medicine
ALBRECHT KOSSEL

in recognition of the contributions to our knowledge of cell chemistry made through his work on proteins, including the nucleic substances.

☆

Literature
PAUL JOHANN LUDWIG HEYSE

as a tribute to the consummate artistry, permeated with idealism, which he has demonstrated during his long productive career as a lyric poet, dramatist, novelist and writer of world-renowned short stories.

☆

Peace
PERMANENT INTERNATIONAL PEACE BUREAU, Bern.

1910

ALBRECHT KOSSEL (1853-1927)

© The Nobel Foundation

The Nobel Prize in Physiology or Medicine was awarded to **Albrecht Kossel** "for contributions to the chemistry of the cell made through his work on proteins, including the nucleic substances."

Albrecht Kossel was born in Rostock, Mecklenburg (now Germany). He studied medicine in the newly founded University of Strasbourg but never practiced. Profoundly influenced by the teachings of Anton de Bary, Waldeyer, Kundt, Baeyer and Felix Hoppe-Seyler (under whom he worked), he devoted his entire career to research in physiological chemistry. In 1883 Emile du Bois-Reymond offered him the directorship of the Chemical Division of the Institute of Physiology in Berlin.

Kossel chose to work on biochemistry, then known popularly as 'physiological chemistry.' By combining advances in organic chemistry with studies in animal physiology, Kossel and his colleagues explored the 'chemistry of life.' In 1879 Kossel began studying the chemical contents of nuclein, a nuclear substance discovered ten years earlier by Friedrich Miescher, another Hope-Seyler student. However, methods

34

available to isolate, purify and analyze the biochemical contents of nuclein were primitive.

In the 1880s Kossel found that upon hydrolysis, nuclein yielded two cleavage products, the already known guanine, and a new product, adenine, discovered by Kossel. By further improving Richard Altmann's techniques of protein purification and isolation, Kossel obtained protein-free nucleins in 1893. Utilizing fish-roe, he studied hexone bases and protamines, and in 1896 discovered histidine. Also, from the nuclein of the thymus gland he isolated two new bases, thymine and cytosine—the chemicals that were shown to be part of the structure of the DNA molecule in the 1950s.

Kossel could not have foreseen the growth of molecular and cell biology that revolutionized medicine by the mid-20th century; yet his contributions were pivotal in directing the attention of biochemists worldwide to focus their research on the cell nucleus. At the academic institutions, Kossel also made another important contribution. He actively sought to secure independent chairs in physiology and biochemistry at German universities, thereby helping these subjects to develop.

Although their original discoveries have undergone major revisions, the pioneering contributions by Miescher, Altmann, and Kossel profoundly influenced the growth of biochemistry as a fundamental discipline in the life sciences.

References

1. Felix K. "Albrecht Kossel." In Great Chemists, edited by Eduard Farber. New York: Interscience, 1961.
2. Florkin, M. A History of Biochemistry. Amsterdam: Elsevier, 1972.
3. Fruton JS. "From Nuclein to the Double Helix." In Molecules and Life. New York: Wiley-Interscience, 1972.
4. Gray GW. "The Mother Molecules of Life." Harper's Magazine 204 (April 1952): 51-61.
5. Kennaway E. "Some Recollections of Albrecht Kossel." Annals of Science 8 (1952): 393-397.
6. Kossel A: The Chemical Composition of the Cell, Harvey Lectures, 1911, pp. 33-51.

1911

☆

Physics
WILHELM WIEN
for his discoveries regarding the laws governing the radiation of heat.

☆

Chemistry
MARIE CURIE (nee Marie Sklodowska)
in recognition of her services to the advancement of chemistry by the discovery of the elements radium and polonium, by the isolation of radium and the study of the nature and compounds of this remarkable element.

☆

Physiology or Medicine
ALLVAR GULLSTRAND
for his work on the dioptrics of the eye.

☆

Literature
COUNT MAURICE (MOORIS) POLIDORE MARIE BERNHARD MAETERLINCK
in appreciation of his many-sided literary activities, and especially of his dramatic works, which are distinguished by a wealth of imagination and by a poetic fancy, which reveals, sometimes in the guise of a fairy tale, a deep inspiration, while in a mysterious way they appeal to the readers' own feelings and stimulate their imaginations

☆

Peace
TOBIAS MICHAEL CAREL ASSER
Cabinet Minister, the Netherlands. Member of the Privy Council. Initiator of the International Conferences of Private Law at the Hague, and
ALFRED HERMANN FRIED
Austria. Journalist. Founder of the peace journal *Die Waffen Nieder.*

1911

Allvar Gullstrand (1862-1930)

© The Nobel Foundation

Allvar Gullstrand received the Nobel Prize in Physiology or Medicine "for his work on the dioptrics of the eye."

Gullstrand was born in Landskrona, Sweden, and studied at the universities at Uppsala, Vienna and Stockholm. He obtained a medical license in 1888 and a doctorate in 1890. His dissertation was entitled "A Contribution to the Theory of Astigmatism." In 1894 he was appointed Professor of Ophthalmology at Uppsala University—the first to hold such a position.

Gullstrand applied the principles of physics and geometry—subjects he had taught himself—to his knowledge of anatomy and physiology and studied the phenomenon of accommodation in the human eye. Accommodation is the physiological word to describe the unique ability of the eye to effortlessly, focus on objects that are far away as well as those that are very nearby.

The conventional theories held that contraction of ciliary muscles caused a change in the anterior curvature of the lens, increasing the refractory power for accommodation. Gullstrand questioned this: unlike a glass lens, he reasoned, the human lens is made of numerous fibers of varying length and composition; therefore, each fiber must affect accommodation. Gullstrand developed models to show that only about two-thirds of the lens refractory power was "extracapsular," gained from an increase in its anterior curvature, and the remaining was "intracapsular," from a re-arrangement of internal elements. This showed that unlike the camera lens, the lens in the human eye undergoes a complex series of changes to enable us to always see precisely and clearly and in all directions, almost instantaneously.

In 1911 Gullstrand invented the slit lamp and the reflex-free ophthalmoscope. Although his original designs have been modified, these basic instruments remain the most enduring of his legacies.

References

1. McGrew, RE, and Margaret P. McGrew. Encyclopedia of Medical History. New York: McGraw-Hill, 1985.
2. Russel, R. "Allvar Gullstrand." In the Who's Who of Nobel Prize Winners, edited by Bernard S. Schlessinger and June H. Phoenix, Ariz: Oryx Press, 1986.
3. Wasson, T, ed. Nobel Prize Winners. New York: H.W. Wilson, 1987.

1912

☆

Physics
NILS GUSTAF DALÉN

for his invention of automatic regulators for use in conjunction with gas accumulators for illuminating lighthouses and buoys.

☆

Chemistry
VICTOR GRIGNARD

for the discovery of the so-called Grignard reagent, which in recent years has greatly advanced the progress of organic chemistry, and

PAUL SABATIER

for his method of hydrogenating organic compounds in the presence of finely disintegrated metals whereby the progress of organic chemistry has been greatly advanced in recent years.

☆

Physiology or Medicine
ALEXIS CARREL

in recognition of his work on vascular suture and the transplantation of blood-vessels and organs.

☆

Literature
GERHART JOHANN ROBERT HAUPTMANN

primarily in recognition of his fruitful, varied and outstanding production in the realm of dramatic art.

☆

Peace
ELIHU ROOT

Former Secretary of State. Initiator of several arbitration agreements.

1912

ALEXIS CARREL (1873-1944)

The Nobel Prize in Physiology or Medicine was awarded to **Alexis Carrel** for his work on "vascular suture and the transplantation of blood vessels and organs."

Carrel was born in Lyons, France. When Alexis was very young his father died, and his mother raised him. After receiving a medical degree from the University of Lyons in 1900, he worked as surgeon and taught anatomy and surgery.

The tragic assassination in 1894 of Sadi Carnot, the President of the French Republic, deeply saddened Carrel. The president had been stabbed, and six surgeons could not stop the bleeding from the torn portal vein, because no safe technique existed for suturing large blood vessels. Carrel became determined to develop techniques for this purpose.

Working in Lyon (1902), Chicago (1904) and the Rockefeller Institute for Medical Research, New York (1906-12), he perfected techniques for repairing torn blood vessels. A classic method he developed was to use three retaining stitches at equidistant points along the torn edge, converting a circular artery into a triangle. He would then stitch each side with artistic precision, establishing re-circulation and more importantly, preventing strictures.

Carrel's surgical skills came in handy in an unusual episode in 1908. A 4-day-old girl with a bleeding disorder was admitted to a New York hospital and her doctors concluded that she needed a blood transfusion—a procedure that had never been done in newborn infants. Since there were no instruments for pediatric transfusions, a "direct transfusion" was the only possible solution.

Surgeon Alexis Carrel from the Rockefeller Institute was consulted; he then carried out an anastomosis between the popliteal vein of this "hopeless-looking baby" with her father's radial artery. No anesthetic was used on either the baby or the father. "Enough blood was allowed to flow into the baby to change her color from pale whiteness to brilliant red...." wrote the baby's doctor Samuel Lambert. The infant survived. Carrel's name is thus etched in the annals of history as the person who carried out the first reported neonatal blood transfusion procedure.

Carrel was also a pioneer transplantation surgeon. He grafted arteries in laboratory animals using rubber tubes and patches from veins. He transplanted thyroid, spleen, ovary, and kidney. One of the nephrectomized dogs transplanted with a single donor kidney lived for 30 months.

An intense perfectionist and a man of many talents, Carrel became unpopular in later years for views on eugenics and for his supposed role in France during World War II. But his 1935 book, *Man, the Unknown* became an international bestseller.

References

1. Bradley J, A. McGehee H. Two Centuries of American Medicine, 1776-1976. Philadelphia: W.B. Saunders, 1976.
2. Carrel A. Treatment of Infected Wounds, 1917 (with Georges Dehelly).
3. Carrel, A: Man, the Unknown, 1935; The Making of Civilized Man, 1937.
4. Carrel, A: The Culture of Organs, 1938 (with Charles A. Lindbergh).
5. Corner, GW. A History of the Rockefeller Institute: 1901-1953. New York: Rockefeller University, 1965.
6. Lambert SW. Melaena neonatorum with report of a case cured by transfusion. *Medical Record*. Vol 73, No 22; May 30, 1908.

1913

Physics
HEIKE KAMERLINGH-ONNES

*for his investigations on the properties of matter at low temperatures
which led, inter alia to the production of liquid helium.*

Chemistry
ALFRED WERNER

*in recognition of his work on the linkage of atoms in molecules by
which he has thrown new light on earlier investigations and opened
up new fields of research especially in inorganic chemistry.*

Physiology or Medicine
CHARLES ROBERT RICHET

in recognition of his work on anaphylaxis.

Literature
RABINDRANATH TAGORE

*because of his profoundly sensitive, fresh and beautiful verse, by which,
with consummate skill, he has made his poetic thought, expressed
in his own English words, a part of the literature of the West.*

Peace
HENRI LA FONTAINE

Belgium. Member of the Belgian Parliament. President of
the Permanent International Peace Bureau, Berne.

1913

CHARLES RICHET (1850-1935)

© The Nobel Foundation

Charles Richet was awarded the Nobel Prize in Physiology or Medicine "for his work on anaphylaxis."

Richet was born in Paris and obtained a medical degree there in 1869. For much of his life he did immunological research at the physiological laboratory in the Faculty of Medicine in Paris.

While on a cruise in the equatorial seas, Prince Albert of Monaco suggested that Richet and others should study the poison from the Portuguese man-of-war, a coelenterate common to those waters. In Paris, Richet prepared a glycerin extract from their tentacles, which produced typical symptoms of poisoning in laboratory animals. Since Portuguese man-of-war was unavailable in France, Richet studied the poison of Actiniae, a different sea anemone. When he injected 0.1 ml/kg of this poison into dogs, no symptoms occurred; but a second injection of 0.005 ml/kg three weeks later caused an unexpected and violent reactions, often leading to death. To describe these phenomena, Richet in 1902 coined the term "anaphylaxis" or "anti-protection."

Developing unexpected reactions to foreign substances, particularly to vaccines, had been well known since the time of Edward Jenner; but it was Richet who established their regularity. Richet also showed that both immunity and hypersensitivity could be transferred from one experimental animal to another.

Richet was unable to comprehend the biological significance and was troubled by the philosophical 'meaning' of anaphylaxis. Agreeing with contemporary scientists, he had always believed that nature created organisms such that the body's biochemical make up protected the organism from foreign enemies. He was troubled by the notion that nature, which created such 'perfect' specimen as the living organisms, also could produce such lethal hypersensitivity reaction as anaphylaxis. To explain this enigma, Richet reasoned that anaphylaxis was nature's way of protecting the entire species at the expense of an individual member of the species.

When foreign proteins 'contaminate' the chemical integrity of the individual, such impurities may be transferred to future generations, thereby threatening the survival of the very species, he reasoned. Therefore, it appeared reasonable that the individual was being 'sacrificed' through anaphylaxis to protect the remaining members of the species. While he did not realize it at the time, this explanation is one way of restating Darwin's theory of natural selection with a philosophical ring.

In his later years, Richet continued to pursue his deep interest in psychology, telepathy, clairvoyance, and aeronautics. He even designed an airplane, wrote several novels and plays, and remained a staunch pacifist. He was active in academics too; he edited *Journal de physiologie et de pathologie générale.* Celebrating half a century of his scientific work, the French Government bestowed upon Richet, the Grand Cross of the Legion of Honor in 1926.

References

1. Glasser RJ. The Body Is the Hero. New York: Random House, 1976.
2. Klein J. Immunology: The Science of Self-Nonself Discrimination. New York: John Wiley & Sons, 1982.
3. Richet C. The Impotence of man, 1929; Au secours! 1935.
4. Richet C. The Story of Civilization through the Ages, 1930.
5. Silverstein AM. A History of Immunology: New York: Academic Press, 1989.

1914

☆

Physics
MAX VON LAUE
for his discovery of the diffraction of X-rays by crystals.

☆

Chemistry
THEODORE WILLIAM RICHARDS
*in recognition of his accurate determinations of the atomic
weight of a large number of chemical elements.*

☆

Physiology or Medicine
ROBERT BÁRÁNY
for his work on the physiology and pathology of the vestibular apparatus.

☆

Literature
No Award:

☆

Peace
No Award

1914

ROBERT BÁRÁNY (1876—1936)

© The Nobel Foundation

Robert Bárány was awarded the Nobel Prize in medicine, "for his work on the physiology and pathology of the vestibular apparatus."

Bárány was born in Vienna and obtained a medical degree in 1900. After studying psychiatry and neurology in Germany, he returned to Vienna in 1903 and began working as an otologist.

It was then a routine practice to treat ear infections by irrigating external auditory canals. Some patients complained of dizziness during this procedure and Bárány noted that the dizzy spells always were accompanied by nystagmus (repeated jerky horizontal movements of the eyes). One patient even said that he felt dizzy only when the water was "not warm enough."

This led Bárány to study the relationship between the temperature of the irrigating fluid, nystagmus and dizziness. He soon established that in healthy persons, the water temperature and head position bore a predictable relationship to the direction of the nystagmus, and the development of dizzy spells. He argued that these must be due to reflexes from the flow of fluids in the semicircular canals (in the inner ear), which must also influence keeping the body in balance against the actions of gravity. Based on these principles he developed the 'caloric test' for diagnosis of neurological and middle ear diseases.

Bárány's contributions to otology came at a time little else was known about either the mechanisms of workings of the inner ear, or the neurophysiology of maintaining body balance. "Bárány caloric test" remains a valuable clinical, non-invasive tool in the diagnosis of vestibular dysfunction.

World War I took its toll on both the scientific career and personal life of Bárány. At the beginning of the War, he joined the army and was captured by the Russians. When the Nobel Prize was announced, Bárány was a prisoner-of-war in a jail in Turkey. It was on December 11, 1916, two years after winning the Nobel Prize, that he was able to deliver his Nobel lecture.

Later in Vienna, Bárány, now a celebrity, faced accusations from colleagues for not acknowledging his collaborators' contributions. Prevailing anti-Semitism made matters worse for Bárány. Although most of the charges were later proved false, Bárány was dejected. He left Vienna and spent the rest of his life working as director of the otolo-rhino-laryngology clinic at the University of Uppsala, Sweden.

References

1. Bárány, R. Physiology and Pathology of the Semicircular Canal, 1910.
2. Haymaker W. ed. The Founders of Neurology: One Hundred and Thirty-Three Biographical Sketches by Eighty-eight Authors. Springfield, Ill. Charles C. Thomas, 1970.
3. McHenry, LC. Garrison's History of Neurology. Springfield, IL. Charles C. Thomas, 1969.

1915

☆

Physics

**SIR WILLIAM HENRY BRAGG and SIR
WILLIAM LAWRENCE BRAGG**

for their services in the analysis of crystal structure by means of X-rays.

☆

Chemistry

RICHARD MARTIN WILLSTÄTTER

for his research on plant pigments, especially chlorophyll.

☆

Physiology or Medicine

No Prize Awarded

☆

Literature

ROMAIN ROLLAND

*as a tribute to the lofty idealism of his literary production and to the sympathy
and love of truth with which he has described different types of human beings.*

☆

Peace

No Awards

1916

Physics
No Award
☆

Chemistry
No Award
☆

Physiology or Medicine
No Award
☆

Literature
CARL GUSTAF VERNER VON HEIDENSTAM
in recognition of his significance as the leading
representative of a new era in our literature.
☆

Peace
No Award

1917

☆

Physics
CHARLES GLOVER BARKLA
for his discovery of the characteristic Röntgen radiation of the elements.

☆

Chemistry
No Award

☆

Physiology or Medicine
No Award

☆

Literature
KARL ADOLPH GJELLERUP
for his varied and rich poetry, which is inspired by lofty ideals, and
HENRIK PONTOPPIDAN
for his authentic descriptions of present-day life in Denmark.

☆

Peace
INTERNATIONAL COMMITTEE OF THE REDCROSS
Geneva.

1918

Physics
MAX KARL ERNST LUDWIG PLANCK
in recognition of the services, he rendered to the advancement of physics by his discovery of energy quanta.

☆

Chemistry
FRITZ HABER
for the synthesis of ammonia from its elements.

☆

Physiology or Medicine
No Award

☆

Literature
No Award

☆

Peace
No Award

1919

✩

Physics
JOHANNES STARK
*for his discovery of the Doppler effect in canal rays and
the splitting of spectral lines in electric fields.*

✩

Chemistry
No Award

✩

Physiology or Medicine
JULES BORDET
for his discoveries relating to immunity.

✩

Literature
CARL FRIEDRICH GEORG SPITTELER
in special appreciation of his epic, Olympian Spring.

✩

Peace
THOMAS WOODROW WILSON
President of the United States of America.
Founder of the League of Nations

1919

JULES BORDET (1870-1961)

© The Nobel Foundation

The Nobel Prize in Physiology or Medicine was awarded to **Jules Bordet** "for his discoveries in regard to immunity."

Jules Bordet was born in Soignies, Belgium. He studied at the University of Brussels and obtained a medical degree in 1892. He worked at the Pasteur Institute in Paris (1894-1901) and later founded the Belgium Pasteur Institute.

From his early immunological studies Bordet showed that two substances were involved in killing the invading bacteria—'bacteriolysis.' One was a chemical substance specifically acting against bacteria, developed by the body in response to immunization, and the other, a thermo-labile substance, generic to all serum. The latter he termed the "alexin" ("*alexo,*" I ward off, Greek). These two substances later came to be known as "antibody" and "complement," respectively.

In 1894 Pfeiffer had shown that the bacteriolytic properties of immune serum can be used to make the diagnosis of cholera in the laboratory. Bordet modified this concept and showed that the procedure can be performed in a test tube. Using this knowledge and working with Octave Gengou, a fellow Belgian, Bordet developed the 'complement fixation test.' This test was later modified and adapted by Wassermann, Neisser, and Bruck in their historic 1906 study of antibody detection in the serum of syphilitic monkeys.

Because of World War-related difficulties, the awards for 1919 and 1920 were announced simultaneously, and presented in December of 1920. At the time of the announcement, however, Bordet was touring in the USA, raising funds to reconstruct the Faculty of Medicine for the Free University of Brussels, which had been severely damaged during the War. With an apology for the absence of the "illustrious laureate" Professor Alfred Patterson of the Royal Caroline Institute made the presentation address in Stockholm on December 10, 1920.

Although officially retired in 1940, Bordet continued to work; collaborating with his son Paul Bordet, he contributed to several scientific theories in bacteriology. A man of deep social consciousness, Bordet was a staunch pacifist. Along with nine thousand other eminent scientists from around the world, Bordet also signed the controversial petition Linus Pauling circulated protesting continued testing of nuclear bombs. (Pauling received the 1954 Nobel Prize for Chemistry and the 1962 Nobel Prize for Peace.)

Besides many contributions to the fundamental concepts of immunology and bacteriology, Bordet and his brother-in-law, Octave Gengou together discovered the pertussis bacillus in 1906, the causative agent of whooping cough—the bacterium was known for a time as the *Bordet-Gengou bacillus*.

References

1. Beumer, J. "Jules Bordet: 1870-1961." Journal of General Microbiology 29 (1962): 1-13.
2. Bordet J: Studies in Immunity, 1909.
3. Most of Bordet's papers are preserved in the Musée Jules Bordet in Brussels. His principal scientific papers were published in either Annales de l'Institut Pasteur or Comptes rendus Société Biologie.

1920

☆

Physics
CHARLES-ÉDOUARD GUILLAUME
in recognition of the service, he has rendered to precision measurements
in Physics by his discovery of anomalies in nickel steel alloys.

☆

Chemistry
WALTHER HERMANN NERNST
in recognition of his work in thermochemistry.

☆

Physiology or Medicine
SCHACK AUGUST STEENBERGER KROGH
for his discovery of the capillary motor regulating mechanism.

☆

Literature
KNUT PEDERSEN HAMSUN
for his monumental work, Growth of the Soil.

☆

Peace
LEÓN VICTOR AUGUSTE BOURGEOIS
Former Secretary of State, President of the Parliament
Council of the League of Nations.

1920

AUGUST KROGH (1874-1949)

© The Nobel Foundation

August Krogh was awarded the Nobel Prize in Physiology or Medicine "for his discoveries of the regulation of the motor mechanism of capillaries."

Krogh was born in Grenaa, Denmark, and studied physiology and zoology at the University of Copenhagen. From 1897 he worked as assistant to the famous physiologist Christian Bohr, who was also Krogh's most important mentor. Krogh married Marie Jørgensen, a physician and scientist, who became his life-long research collaborator.

During an expedition in Greenland in 1902, Krogh measured gas tensions in waters from different sources and discovered that the oceans affected the atmospheric carbon dioxide. Later, in collaboration with Bohr and Hasselbach, Krogh used his "micro tonometer," an instrument for measuring minute quantities of gas tensions in air bubbles and showed that blood oxygen affinity depended upon its carbon dioxide tension. He also proved that in the lungs, gas exchange occurred through diffusion, not secretion—a theory championed by

Bohr. Krogh's refutation of oxygen secretion theory led to a permanent rift between him and Bohr.

From 1911 Krogh worked in his own laboratory and made additional discoveries concerning capillary circulation and muscle physiology—the work that would lead him to the Nobel Prize. He described the "capillo-motor-mechanism," or the contraction and expansion of capillaries and arterioles in muscles and other tissues and deduced that this occurred through chemical mediators.

A man of insatiable curiosity, Krogh continued to carry out much valuable research throughout his life. He was devoted to the study of physiology of respiration in aquatic animals; in 1935 he published a landmark study describing the role of isotopes in biological studies. After World War II ended, he resigned his professorship at the university, but continued his work in his private laboratory, studying energy metabolism, and developed an ingenious, miniature wind tunnel to study the flight of grasshoppers.

Krogh remains one of the most influential physiologists of the 20th century. His books, *The Respiratory Gas Exchange of Animals and Men* (1916) and *The Anatomy and Physiology of Capillaries* (1922), remain classics.

References

1. Dejours P. "August Krogh and the Physiology of Respiration." Scandinavian Journal of Respiratory Disease 56 (1975): 337-345.
2. Krogh A: Frøernes Hud-og Lungerespiration, 1903
3. Krogh A: Kortfattet laerebog I mennshets Fysiologi, 1908
4. Krogh A: The Respiratory Exchange of Animals and Men, 1916
5. Krogh A: The Anatomy and Physiology of Capillaries, 1922
6. Krogh A: Osmotic Regulation in Aquatic Animals, 1939
7. Krogh A: The Comparative Physiology of Respiratory Mechanisms, 1941.

1921

☆

Physics
ALBERT EINSTEIN

*for his services to Theoretical Physics, and especially for his
discovery of the law of the photoelectric effect.*

☆

Chemistry
FREDERICK SODDY

*for his contributions to our knowledge of the chemistry of radioactive
substances, and his investigations into the origin and nature of isotopes.*

☆

Physiology or Medicine
No Award: No consensus reached.

☆

Literature
ANATOLE FRANCE (penname JACQUES ANATOLE THIBAULT)

*in recognition of his brilliant literary achievements, characterized as they are by a
nobility of style, a profound human sympathy, grace, and a true Gallic temperament.*

☆

Peace
KARL HJALMAR BRANTING

Prime Minister. Swedish Delegate to the Council of the League of Nations, and

CHRISTIAN LOUS LANGE

Secretary General of the Inter-Parliamentary Union, Brussels.

1922

Physics
NIELS BOHR

*for his services in the investigation of the structure of atoms
and of the radiation emanating from them.*

Chemistry
FRANCIS WILLIAM ASTON

*for his discovery, by means of his mass spectrograph, of
isotopes, in a large number of non-radioactive elements,
and for his enunciation of the whole-number rule.*

Physiology or Medicine
SIR ARCHIBALD VIVIAN HILL

for his discovery relating to the production of heat in the muscle, and

OTTO FRITZ MEYERHOF

*for his discovery of the fixed relationship between the consumption
of oxygen and the metabolism of lactic acid in the muscle.*

Literature
JACINTO BENAVENTE Y MARTÍNEZ

*for the happy manner in which he has continued the
illustrious traditions of the Spanish drama.*

Peace
FRIDTJOF NANSEN

Norway. Scientist. Explorer. Norwegian Delegate to the League
of Nations; Originator of the Nansen passports for refugees.

1922

SIR ARCHIBALD VIVIAN HILL (1886-1977) AND OTTO FRITZ MEYERHOF (1884-1951)

Archibald Hill   **Otto Fritz Meyerhof**
© The Nobel Foundation

Sir Archibald Vivian Hill (left) and **Otto Fritz Meyerhof** received the Nobel Prize in Physiology or Medicine for work related to muscle energy metabolism.

Hill was born in Bristol and studied mathematics and physiology at the Trinity College, Cambridge. A pioneer biophysicist, Hill improved galvanometers and developed sensitive equipment for temperature measurements. He showed that the heat produced from muscle work was both from its contraction and relaxation, in proportion to lactic acid production and removal.

Otto Fritz Meyerhof was born in Hanover and graduated in medicine from Heidelberg University. Influenced by Otto Warburg, who was to win the Nobel Prize in 1931, Meyerhof worked on cell physiology and biochemistry at the University of Kiel. He used precise laboratory methods to the study of biochemistry of muscle contraction and energy metabolism.

With impeccable accuracy, Meyerhof showed that one of four molecules of lactic acid formed during muscle contraction was oxidized and the remainder reverted to carbohydrates. Within four years, he had deciphered almost the entire glycolytic pathway, complementing Hill's research. Remarkably, both scientists had used isolated frog muscles in their studies—the classical frog muscle preparation. Meyerhof would later say that Hill's muscle heat measurements served as the "beacon" pointing a way to the understanding of his own research findings.

In 1938 Meyerhof fled Nazi Germany to France, Spain, and later to the USA. A philosopher-scientist, stimulating conversationalist, and an accomplished painter, Meyerhof had an enormous influence on the growth of biochemistry in this century.

Hill continued to study muscle physiology in humans. From his work on exercise physiology, he coined the term "oxygen debt" to describe anaerobic muscle work. He also showed that small, but significant amounts of heat are produced during transmission of nerve impulses.

Besides his scientific contributions, Hill also held prominent roles in public life. He fought for the intellectual freedom of scientists and sought to eliminate parochialism in research funding. He served as a Member of Parliament (1940-1945) representing the Cambridge University in the House of Commons, a member of the University Grants Committee (1937-1944), and served on the Science Committee of the British Council (1946-1956). In the 1940s, he advised the Indian Government on its reconstruction efforts in matters related to science.

References

1. Gemmil, CL. Recollections of Professor Otto Meyerhof. Virginia Medical College, Richmond, Quarterly 2, no. 2 (1966): 141-142.
2. Hill AV: The Heat of Shortening and the Dynamic Constants of Muscle. Proceedings of the Royal Society of London, series B, vol. 126, 1938
3. Hill AV: The Thermodynamics of Muscle. Brit Med Bul vol. 12, 1956
4. Hill AV: Muscular Activity, 1926
5. Hill AV: The Ethical Dilemma of Science and Other Writings. London, Scientific Book Guild; 1962.
6. Hill AV: The Heat Production of Muscle of Nerve. Annual Review of Physiology, vol. 21, 1959, pp. 1-18
7. Matthews GG. Cellular Physiology of Nerve and Muscle, Palo Alto, CA, Blackwell Scientific Publications, 1986.
8. Meyerhof OF. Zur Energetik der Zellvorgänge, 1914 (Chemical Dynamics of Life Phenomena, 1924).

1923

☆

Physics
ROBERT ANDREWS MILLIKAN
*for his work on the elementary charge of electricity
and on the photoelectric effect.*

☆

Chemistry
FRITZ PREGL
for his invention of the method of micro-analysis of organic substances.

☆

Physiology or Medicine
SIR FREDERICK GRANT BANTING and
JOHN JAMES RICHARD MACLEOD
for the discovery of insulin.

☆

Literature
WILLIAM BUTLER YEATS
*for his always inspired poetry, which in a highly artistic
form gives expression to the spirit of a whole nation.*

☆

Peace
No Award

1923

SIR FREDERICK G. BANTING (1891-1941) AND JOHN J. R. MACLEOD (1876-1935)

Frederick Banting   **John Macleod**
© The Nobel Foundation

Frederick Banting (left) and **John Macleod** shared the Nobel Prize in Physiology or Medicine for discovery of insulin.

Banting was born in Alliston, Ontario, Canada. He served in the Canadian army briefly before beginning a practice in London, Ontario.

Although a deficiency of a pancreatic hormone, possibly from the islets of Langerhans, was believed to cause diabetes, pancreatic extracts had proved ineffective as a treatment. Banting read a paper by Moses Barron showing that a ligation of the pancreatic duct caused the gland to atrophy, preserving the islets. Banting thought of repeating those studies to obtain pure islet cell extracts and approached John Macleod for help in 1920. Professor Macleod, a Scotsman, had been Chief of Physiology at the University of Toronto since 1918.

Macleod was initially reluctant but agreed to provide a laboratory and arranged for Charles Best, an undergraduate seeking a research job, to assist Banting. While Macleod vacationed in Scotland in the summer of 1921, Banting and Best obtained an extract from dog pancreatic islet cells and demonstrated its powerful anti-diabetic property. They named the extract "isletin," but as per Macleod's suggestion, renamed it "insulin," from 'insula' for island in Latin. (Sharpey-Schafer had coined the term for a hypothetical islet cell hormone). The news of the first successful clinical testing of insulin on January 23, 1922, by Banting and Best was an international sensation.

Critical of Nobel Foundation's omission of Charles Best in the Prize, Banting gave one-half of his share of the prize money to Best. Macleod gave a portion of *his* share to James Collip, the biochemist who had helped in insulin extraction.

Banting was knighted in 1934; he died in a plane crash during the Second World War. In 1928 Macleod returned to Aberdeen, where he died of arthritis in 1935.

References

1. Banting FG. (with C.H. Best). Insulin in Treatment of Diabetes Mellitus. Journal of Metabolic Research, vol. 2, 1922, pp. 5547-604.

2. Banting FG. Effect of Pancreatic Extract (Insulin) on Normal Rabbits. American Journal of Physiology, vol. 62, 1922, pp. 162-176

3. Banting FG. Internal Secretion of Pancreas. Journal of Laboratory and Clinical Medicine, vol. 7, 1922, pp. 251-326

4. Bliss M. Banting: A Biography. Toronto: McClelland and Stewart, 1984. Bliss M. The Discovery of Insulin. Chicago: University of Chicago Press, 1982.

5. Harris S. Banting's Miracle: The Story of the Discovery of Insulin. Philadelphia: J.B. Lippincott, 1946.

6. Wrenshall, G.A., Hetenyi, and W.R. Feasby. The Story of Insulin: Forty Years of Success Against Diabetes. London: Bodley Head, 1962.

1924

WILLEM EINTHOVEN (1860-1927)

© The Nobel Foundation

Willem Einthoven was awarded the Nobel Prize in Physiology or Medicine for his discovery of the mechanism of electrocardiography.

Einthoven was born in 1860 in Semarang, Java, Dutch East Indies (Indonesia). His father died when Willem was ten, and the family returned to Holland. At age 25 Einthoven received MD and PhD degrees from the University of Utrecht and the same year he was appointed professor at the University of Leiden.

Einthoven applied fundamental principles of physics in solving physiological problems. While in London for a conference in 1887, he witnessed Augustus Waller of St. Mary's Hospital demonstrate an "electrogram"— recordings of waveforms from a man's heart with a capillary electrometer. After seven subsequent years of intense work to develop a similar instrument capable of recording minute electrical currents, Einthoven eventually invented the string galvanometer with which he recorded the first full-fledged "electrocardiogram"— a term he coined.

Einthoven also standardized ECG gain setting, paper speed, connections for leads (I—III), and the nomenclature, "P, Q, R, S, and T." With the string galvanometer, he recorded electrical changes in the retina, vagus nerve, and sympathetic ganglion; but electrocardiography remains his enduring legacy.

A keen sportsman, Einthoven founded the Utrecht Student Rowing Club. He stressed the importance of physical fitness and urged colleagues "not to let the body perish." When the Nobel Prize was announced, Einthoven was on a lecture tour in the United States, and hence unable to attend the scheduled ceremonies in Stockholm. That year's physics laureate Karl Siegbahn, and the literature laureate Wladslaw Reymont also could not go to Stockholm, and hence all traditional Nobel ceremonies were canceled. Einthoven delivered his Nobel lecture on December 11, 1925, one year after the scheduled date.

References

1. Burch GE, De Pasquale NP. A History of Electrocardiography. Chicago: Year Book Medical Publishers, 1964.
2. Cooper JK. "Electrocardiography One Hundred Years Ago." The New England Journal of Medicine 315 (August 14, 1986); 461-464.
3. Einthoven W: The Form and Magnitude of the Electrical Response of the Eye to Stimulation by Light at Various Intensities," Quarterly Journal of Experimental Physiology, vol. 1, 1908, pp. 373-416 (with W.A. Jolly).

1925

☆

Physics
JAMES FRANCK and **GUSTAV HERTZ**
for their discovery of the laws governing the
impact of an electron upon an atom.

☆

Chemistry
RICHARD ADOLF ZSIGMONDY
for his demonstration of the heterogenous nature of colloid
solutions and for the methods he used, which have since
become fundamental in modern colloid chemistry.

☆

Physiology or Medicine
No Award: No consensus reached

☆

Literature
GEORGE BERNARD SHAW
for his work which is marked by both idealism and humanity, its
stimulating satire often being infused with a singular poetic beauty.

☆

Peace
SIR AUSTEN CHAMBERLAIN
Foreign Minister. Negotiator of the Locarno Treaty, and
CHARLES GATES DAWES
Vice-President of the United States of America. Chairman of the
Allied Reparation Commission. Originator of the Dawes Plan.

1926

☆

Physics
JEAN BAPTISTE PERRIN
*for his work on the discontinuous structure of matter, and
especially for his discovery of sedimentation equilibrium.*

☆

Chemistry
THEODOR SVEDBERG
for his work on disperse systems.

☆

Physiology or Medicine
JOHANNES ANDREAS GRIB FIBIGER
for his discovery of the Spiroptera carcinoma.

☆

Literature
GRAZIA DELEDDA (penname of GRAZIA MADESANI)
*for her idealistically inspired writings which with plastic
clarity picture the life on her native island and with depth
and sympathy deal with human problems in general.*

☆

Peace
ARISTIDE BRIAND
Foreign Minister; Negotiator of the Locarno
Treaty and the Briand-Kellogg Pact, and
GUSTAV STRESEMANN
Former Lord High Chancellor (Reichs-kanzler). Foreign
Minister. Negotiator of the Locarno Treaty.

1926

JOHANNES ANDREAS GRIB
FIBIGER (1867—1928)

© The Nobel Foundation

Johannes Andreas Grib Fibiger was awarded the Nobel Prize in Physiology or Medicine for discovering *Spiroptera carcinoma*.

Fibiger was born in Silkeborg, Denmark. He studied bacteriology under Robert Koch and in 1900 became Director at the Copenhagen's Institute of Pathological Anatomy.

In 1907 Fibiger discovered papillomatous stomach tumors in laboratory mice. Within the tumors he found peculiar nematodes, which he later named *Spiroptera neoplastica*. Regarding nematodes, the cancer-causing agents, Fibiger pursued his studies but failed to find any parasite in nearly 1000 mice tested. Later in a Copenhagen sugar factory infested with mice and cockroaches, he found the same nematodes in 75% of mice, 20% of which also had tumors. In 1913 he proposed that cockroaches get infested by eating mice excreta containing parasitic eggs, and the mice get re-infested by eating the larvae-laden cockroaches; irritation or chemicals from the nematodes caused cancers.

Some experts agreed, but most were skeptical of Fibiger's hypothesis.

They could not replicate his findings and concluded that parasites were incidental findings. Later, diets deficient in vitamin A were also shown to cause mice stomach cancers, and Fibiger's theory was forgotten. Because of the 'error' of honoring Fibiger, the Nobel Foundation avoided recognizing subsequent cancer research for forty years.

There is little dispute today about infections causing malignancies. The recent discovery of an association between *Helicobacter pylori* and stomach cancer has further enhanced interest in Fibiger. Whether what Fibiger saw was *H. pylori* has not been deciphered.

Despite considerable criticism Fibiger received during his lifetime and after his death, few if any doubt that he was an eminent scientist deserving our respect and recognition. Dr. William Campbell, a parasitologist in the USA said recently: "Fibiger may have been barking up the wrong tree, but he was still a great Dane."

References

1. Fibiger J. "On the Transmission of Spiroptera Neoplastica to the Rat as a Method of Producing Cancer Experimentally, 1918.
2. Fibiger J. Investigations on the Relationship of Human and Bovine Tuberculosis and Tubercle Bacilli, 1908
3. Fibiger J. Investigations on the Spiroptera Cancer III, 1918.
4. Fibiger J. Investigations upon Immunization Against Metastasis Formation in Experimental Cancer," 1927.
5. Fibiger J. On Spiroptera carcinomata and Their Relation to True Malignant Tumors," 1919
6. Hitchcock CR, Bell ET. Studies on the Nematode Parasite, Gongylonema neoplasticum. Journal of the National Cancer Institute 12 (June 1952): 1345-1387.
7. Oberling C. The Riddle of Cancer. New Haven, Conn.: Yale University Press, 1952.

1927

☆

Physics
ARTHUR HOLLY COMPTON
for his discovery of the effect named after him, and
CHARLES THOMSON REES WILSON
for his method of making the paths of electrically charged particles visible by condensation of vapor.

☆

Chemistry
HEINRICH OTTO WIELAND
for his investigations of the constitution of the bile acids and related substances.

☆

Physiology or Medicine
JULIUS WAGNER-JAUREGG
for his discovery of the therapeutic value of malaria inoculation in the treatment of dementia paralytica.

☆

Literature
HENRI BERGSON
in recognition of his rich and vitalizing ideas and the brilliant skill with which they have been presented.

☆

Peace
FERDINAND BUISSON
Former Professor at the Sorbonne University, Paris and President of the League for Human Rights, and
LUDWIG QUIDDE
Historian; Professor at Berlin University. Member of Germany's constituent assembly 1919.

1927

Julius Wagner-Jauregg (1857-1940)

© The Nobel Foundation

Julius Wagner-Jauregg was awarded the Nobel Prize in Physiology or Medicine "for discovering the therapeutic value of malaria for dementia paralytica."

He was born Julius Wagner in Wels, Austria. The hyphenated "Wagner-Jauregg" replaced "von-Jauregg," his father's title of nobility that had been discontinued after the First World War. Wagner-Jauregg studied medicine in Vienna and later, without training, began work as a psychiatrist. About the choice of this specialty, he once remarked, "it harmed neither myself nor psychiatry."

Untreated syphilis often led to the dreaded *dementia paralytica*, general paralysis of the insane (GPI). In 1887, Wagner-Jauregg noted that when patients with GPI developed fevers from other causes, their mental status improved, sometimes for years. After long contemplation and analyses of the available data, he felt convinced that 'fever therapy' might work in neurosyphilis.

In 1917 Wagner-Jauregg saw a patient with a mild case of tertian malaria. He injected that patient's blood into nine GPI patients, all of whom developed malaria as predicted. On follow-up, three were 'cured' of GPI, three improved slightly, two were unchanged, and one had died of malaria. Encouraged by these results, Wagner-Jauregg continued 'malaria therapy' for GPI and reported 83% remission. After initial reluctance, other experts followed suit. Thousands of patients were treated the world over, and until the advent of penicillin, fever therapy remained the mainstay for GPI.

The first of the two psychiatrists to receive a Nobel Prize thus far, * Wagner-Jauregg remains controversial, but we must regard his work in the context of contemporary science. Neurosyphilis was dreadful and incurable, and the prevailing notion was that desperate maladies justify desperate remedies—a feeling not much different from today. Wagner-Jauregg considered mental illnesses as biological phenomena for which he sought 'physical' cures. In this regard, his views seem contemporary—although not his remedies.

(*Eric Kandel, who shared the 2000 Nobel Prize in Physiology or Medicine, practiced psychiatry for a while.)

References

1. De Kruif, P. Men Against Death. New York: Harcourt, Brace & World, 1932.
2. Kraepelin, E. One Hundred Years of Psychiatry. New York: Citadel Press, 1962.
3. Shorter E. In: *A History of Psychiatry: From the Era of the Asylum to the Age of Prozac.* New York, NY: Wiely Wiley, 1997.
4. Valenstein ES. In: *Great and Desperate Cures: The Rise and Decline of Psychosurgery and other Radical Treatments for Mental Illness.* New York, NY: Basic Books; 1986.

5. Wagner-Jauregg J. "Über die Einwirkung der Malaria auf die progressive Paralyse," Psychiatrisch-Neurologische Wochenschrift, vol. 20, 1918-1919, pp. 132-134.

1928

Physics

SIR OWEN WILLANS RICHARDSON

*for his work on the thermionic phenomenon and especially
for the discovery of the law named after him.*

☆

Chemistry

ADOLF OTTO REINHOLD WINDAUS

*for the services rendered through his research into the constitution
of the sterols and their connection with the vitamins.*

☆

Physiology or Medicine

CHARLES JULES HENRI NICOLLE

For his work on typhus.

☆

Literature

SIGRID UNDSET

for her powerful descriptions of northern life during the Middle Ages.

☆

Peace

No Award

1928

CHARLES JULES HENRY NICOLLE (1866-1936)

© The Nobel Foundation

Charles Jules Henry Nicolle was awarded the Nobel Prize in Physiology or Medicine "for his work on typhus."

Nicolle was born in Rouen, France. Literature was his first love but instead he chose medicine as per his father's wish. After studies in Rouen and Paris, in 1902 he became the director of the Pasteur Institute in Tunis, which he transformed into a world-renowned laboratory for vaccine and bacteriology research.

The causative agent of epidemic typhus, *Rickettsia prowazekii*, had been discovered by Howard Rickettsia and Stanislaus Prowazek, both of whom had died from typhus. In 1909 Nicolle observed that although the typhus patients were highly infectious outside the hospital, soon after receiving a hot bath and fresh clothing inside the hospital they ceased to be contagious. Suspecting body lice to be typhus vectors, Nicolle conducted a series of epoch-making experiments.

By injecting a typhus patient's blood, Nicolle induced typhus in a chimpanzee. Then he transferred some lice from the body of typhus-infested chimpanzee to macaques, all of which developed typhus. He concluded that typhus was propagated through lice bites as well as their excreta. Later, Nicolle developed a vaccination for typhus. These discoveries helped in typhus-eradication measures during the two World Wars.

In 1908 Nicolle and associates discovered *Toxoplasma gondii* in a Tunisian rodent; later, they developed vaccinations for Malta, tick and scarlet fevers. Nicolle's innovative culture techniques greatly aided in the study *Leishmania donovani* and *tropica*.

A philosopher and poet, Nicolle was loved and admired by all who knew him. Between 1930 and 1934 he wrote five books related to infections, biology, and the future of mankind. In all his talks and writings Nicolle reflected that as man conquers old infections new and unforeseen ones might emerge. Thus, many historians see his views as prophetic predictions about such 'new' illnesses as acquired immune deficiency syndrome (AIDS), prion-mediated disorders, and other rare forms of viral and protozoal infections.

References

1. Budd W. On the Causes of Fevers (1839). Edited by Dale C. Smith. Baltimore: The Johns Hopkins University Press, 1984.
2. Bulloch W. The History of Bacteriology. New York: Oxford University Press, 1938.
3. Nicolle C. Recherches experimentales sur le typhus exanthématique. Annales de l'Institut Pasteur, vol. 24, 1910, pp. 243-275; vol. 25, 1911, pp. 97-144; vol. 26, 1912, pp. 250-280 and 332-335.
4. Zinsser H. Rats, Lice, and History. Boston: Atlantic Monthly Press, 1935.

1929

☆

Physics
PRINCE LOUIS-VICTOR DE BROGLIE
for his discovery of the wave nature of electrons.

☆

Chemistry
SIR ARTHUR HARDEN and HANS VON EULER-CHELPIN
for their investigations on the fermentation of sugar and fermentative enzymes.

☆

Physiology or Medicine
CHRISTIAAN EIJKMAN
for his discovery of the antineuritic vitamin and
SIR FREDERICK GOWLAND HOPKINS
for his discovery of the growth-stimulating vitamins.

☆

Literature
THOMAS MANN
for his great novel, **Buddenbrooks,** *which has won steadily increased recognition as one of the classic works of contemporary literature.*

☆

Peace
FRANK BILLINGS KELLOGG
Former Secretary of State, Negotiated the Briand-Kellogg Pact.

1929

CHRISTIAAN EIJKMAN (1858-1930) AND SIR FREDERICK HOPKINS (1861-1947)

Christiaan Eijkman   **Sir Frederick Gowland Hopkins**
© The Nobel Foundation

Christiaan Eijkman (left) and **Frederick Hopkins** shared the Nobel Prize in Physiology or Medicine for their work related to vitamins.

Hopkins was born in Eastbourne, England, and studied at the University of London and at Guy's Hospital Medical School. In 1902, he joined Trinity College, Cambridge, and from 1914, he became the head of the department of biochemistry there.

Hopkins inaugurated a new era in biochemistry. He discovered tryptophan, glutathione and xanthine oxidase, and made contributions in protein and uric acid biochemistry. His work showed that animals fed only on amino acids, carbohydrates, fats and salts, developed severe growth failures. In an historic publication of 1912, he described how even small amounts of milk restored growth. He concluded that natural foods contained "essential, growth-promoting, accessory factors', substances later identified as vitamins.

Eijkman was born in Nijkek, The Netherlands and worked as an army doctor. In 1886, the Dutch government deputed him and a colleague to go back to Java to discover the cause of rampant beriberi there. The team's initial research failed, and Eijkman's colleague returned to Holland, while Eijkman stayed in Jakarta. Because chickens were cheap and they contracted beriberi-like illness, Eijkman choose then as experimental animals to find the mysterious "beriberi-causing germs." He injected the "infected" materials from beriberi animals to healthy chickens but could not transfer the infection.

In 1889, a major epidemic of beriberi killed nearly three-fourths of all the animals, including the non-injected controls. However, just as mysteriously, the epidemic vanished suddenly. Eijkman soon learned that a servant had used table scrapings of rice from army officers' mess to feed the chickens, about three weeks before the beriberi epidemic. Later, a new cook had "refused to let the military rice be fed to the civilian chickens," and had switched to the cheaper unpolished rice— which seemed to have cured the chickens.

Eijkman recognized that the 'military rice' had caused hen's beriberi, and the unpolished rice had cured it. First, he thought that the rice polishing had a factor that countered some 'toxins' in the polished rice. He then experimented on some prisoners: by feeding polished rice and then switching back to unpolished rice, he showed that the former diet caused beriberi and the latter cured it. Eijkman surmised that raw rice had a 'anti-beriberi factor,' destroyed by polishing.

At about the same time, Casimir Funk (1884-1967), a Polish-born biochemist in London was studying the chemistry of rice polishing. He isolated a compound and named it thiamin. Using it he cured beriberi in laboratory pigeons. In 1912, Funk coined the term 'vitamine' because he thought thiamin and similar food chemicals were chemically "amines," and were "vital" for survival.

Thus, by the early 1920s, the stage was set to crack the mystery of the ancient nutritional diseases. Eijkman's anti-beriberi factor was shown to be thiamin Funk had discovered; and "vitamine" was shown to be

Hopkin's "essential accessory factor." Since not all these compounds were 'amines', the letter 'e' was dropped, and the generic term 'vitamin' began to be used for all organic micronutrients the body was unable to make. By the mid-century, some 20 vitamins had been discovered, and multi-billion-dollar industries were booming the world over.

Funk also isolated niacin but did not realize its full nutritional value. Three years later he sailed to the USA and spent the rest of his illustrious career in New York. Although Funk was not named in the Nobel Prize, he received many honors; an institute of research was named after him and a Polish postage stamp was issued in his honor in 1992.

Thus, because of surprising discoveries by a Dutch army officer, a British biochemist, and a Polish scientist, most of the vitamin deficiencies known since the time of the Pharaohs, have all but disappeared.

References

1. Bailey, H: The Vitamin Pioneers. Emmaus, Pa. Rodale Books, 1968.
2. Eijkman C. Over gezondheid en ziekte in heete gewesten, 1898.
3. Florkin, M. A History of Biochemistry. Amsterdam: Elsevier, 1972.
4. Harris, LJ. "The Discovery of Vitamins." In: The Chemistry of Life: Eight Lectures on the History of Biochemistry, edited by Joseph Needham. Cambridge, England: Cambridge University Press, 1970.
5. Hopkins F: "A Contribution to the Chemistry of Proteids: Part I, A Preliminary Study of a Hitherto Undescribed Product of Tryptic Digestion," Journal of Physiology, vol. 27, 1901, p. 418 (with S.W. Cole)
6. Hopkins F: Feeding Experiments Illustrating the Importance of Accessory Factors in Normal Dietaries," Journal of Physiology, vol. 44, 1912, p. 425.
7. Leiceser, HM. Development of Biochemical Concepts from Ancient to Modern Times. Cambridge, Mass.: Harvard University Press, 1974.

8. McCollum, EV. A History of Nutrition: The Sequence of Ideas in Nutrition Investigations. Boston: Houghton Mifflin, 1957.
9. Needham J, Baldwin E. Hopkins and Biochemistry. Cambridge, England: W. Heffer, 1949.

1930

☆

Physics
SIR CHANDRASEKHARA VENKATA RAMAN
for his work on the scattering of light and for the discovery of the effect named after him.

☆

Chemistry
HANS FISCHER
for his research into the constitution of haemin and chlorophyll and especially for his synthesis of haemin.

☆

Physiology or Medicine
KARL LANDSTEINER
for his discovery of human blood groups.

☆

Literature
SINCLAIR LEWIS
for his vigorous and graphic art of description and his ability to create, with wit and humour, new types of characters.

☆

Peace
LARS OLOF NATHAN (JONATHAN) SÖDERBLOM,
Archbishop. Leader of the ecumenical movement

1930

KARL LANDSTEINER (1868-1943)

© The Nobel Foundation

Karl Landsteiner was awarded the Nobel Prize in medicine "for his discovery of the human blood groups."

Landsteiner was born in Vienna. There he studied medicine and joined the university as pathologist. Intolerable living conditions in Vienna following the First World War forced the Landsteiner family to take "the last train out" to Holland in 1919. Landsteiner moved to New York in 1922 and joined the Rockefeller Institute.

It was known that serum from one species of animals agglutinated (clumped) red blood cells (RBCs) belonging to other species. Paul Ehrlich had shown that occasionally a weaker, but similar, reaction occurred with serum and RBCs from the members of the same species. Landsteiner was surprised to find that a strong agglutination occurred when human serum was mixed with RBCs from other individuals;

this he discovered using blood from himself and from his laboratory assistants. After much testing, in 1901 Landsteiner proposed that people can be classified into A and B, based on the presence in the blood of specific antibody, and C or "zero," for a group that contained neither A nor B antigens; the *zero* was mistakenly read as 'O' thus the name stuck. His colleagues added the fourth, the "AB" group the next year. Initially, the knowledge of blood groups helped to establish or rule out disputed paternity, but slowly, its implications for transfusion therapy began to be realized.

When he was 72 (in 1940), Landsteiner and his coworker Alexander Wiener made another monumental discovery. Having discovered that antibodies produced by rabbits in response to the injected RBCs from rhesus monkeys cross-reacted with human RBCs, they made a brilliant proposition: The well-known, unexpected reactions that occurred in individuals even after receiving ABO-compatible blood transfusions were probably due to a lack of certain 'common' antigen also found in the monkeys. Landsteiner and Weiner proposed that the rhesus monkey factor was found in about 85% of individuals; Landsteiner called it the "Rhesus Factor (Rh factor)."

However, 10 months earlier in June 1939, Philip A Levine, one of Landsteiner's former associates (who had co-discovered the M, N, and P blood groups with Landsteiner), also had come to a similar conclusion. Based on considerable research using the blood from a patient with transfusion reactions, he had discovered the existence of the rare factor as cause of unexpected intra-group blood transfusion reactions. But Levine had not proposed any name for the new antigen.

These events led to one of the most famous and protracted feuds in medicine about the priority for a discovery. Levine and Weiner bitterly fought privately and publicly. It is widely believed that both were short-listed for a Nobel Prize, but to avoid an open battle between the laureates, the Foundation stayed away from honoring them for their role in the discovery of the Rh factor.

Landsteiner also discovered 'haptens,' the chemicals responsible for the Wassermann reaction, and showed that poliomyelitis could be transmitted to monkeys through spinal cords extracts obtained from polio victims—an important concept aiding in future poliomyelitis research. A quiet and reserved man, Landsteiner was an untiring worker, and according to some of his students, a "hard taskmaster." He suffered a heart attack while working in the laboratory with a pipette in hand. Two days later he died at 75.

References

1. Dixon, Bernard. "Of Different Bloods." Science '84 5 (November 1984): 65-67

2. Landsteiner K. An Agglutinable Factor in Human Blood Recognized by Immune Sera for Rhesus Blood, Proceedings of the Society for Experimental Biology and Medicine, vol. 43, 1940, pp. 223-224 (with Alexander S. Wiener)

3. Landsteiner K. On Agglutination Reactions of Normal Human Blood, Vienna Clinical Weekly, vol. 14, 1901, 1132-1134.

4. Landsteiner K. On the Presence of M Agglutinogens in the Blood Monkeys," Journal of Immunology, vol. 33, 1937, pp. 19-25.

5. Wintrobe MM. Blood, Pure and Eloquent: A Story of Discovery, People, and Ideas. New York: McGraw-Hill, 1980.

6. Zimmerman, DR. Rh: An Intimate History of a Disease and its Conquest. New York, MacMillan Publishing, 1973.

1931

Physics
No Awards

☆

Chemistry
CARL BOSCH and FRIEDRICH BERGIUS
*in recognition of their contributions to the invention and
development of chemical high-pressure methods.*

☆

Physiology or Medicine
OTTO HEINRICH WARBURG
for his discovery of the nature and mode of action of the respiratory enzyme.

☆

Literature
ERIK AXEL KARLFELDT
for his poetry

☆

Peace
JANE ADDAMS
Sociologist. International President of the Women's
International League for Peace and Freedom, and
NICHOLAS MURRAY BUTLER
President of Columbia University. Promoter of the Briand-Kellogg Pact.

1931

OTTO HEINRICH WARBURG (1883—1970)

© The Nobel Foundation

Otto Heinrich Warburg was awarded the Nobel Prize in Physiology or Medicine "for his discovery of the nature and mode of action of the respiratory enzyme."

Warburg was born in Freiburg, Baden, where his father was a prominent physicist. Otto studied chemistry under Emil Fischer in Berlin; later he went to medical school at Heidelberg University and earned his degree in 1911. He was wounded serving in the Prussian Horse Guards during the First World War, for which he received the Iron Cross decoration. Warburg did most of his research at Berlin's Kaiser-Wilhelm Institute, which he renamed the Max Planck Institute for Cell Physiology in 1953.

Warburg's research interest centered on cell energy metabolism. He improved the Haldane-Barcroft blood gas manometer (later called Warburg manometer) to study oxygen consumption in normal and cancerous cells, and photosynthesis in plant tissues. In 1926 Warburg and colleagues discovered that iron containing catalytic pigment, ferment iron, was a key factor in 'cell respiration'— a phrase Warburg coined and the concept he discovered. His methods of studying cell metabolism using thin tissue slices in buffered solutions were widely adopted. In 1944 the Nobel Foundation considered awarding him an unprecedented, second Nobel Prize. Nobel laureates Otto Meyerhof (1921) and Sir Hans Adolf Krebs (1953) were among his illustrious pupils.

Despite his Jewish ancestry, Warburg survived the Nazi regime, perhaps because of his status, connections, and war records. Many scientists including Albert Einstein criticized his tolerance of Hitler. Later in life Warburg came to some disrepute because of his intolerant attitude towards scientists opposing his theories. But Warburg's eminence as one of the most influential scientists of this century remains undisputed.

References

1. Keilin D. The History of Cell Respiration and of Cytochrome, Cambridge, England: Cambridge University Press, 1966.
2. Krebs H. Otto Heinrich Warburg. In Biographical Memoirs of Fellows of the Royal Society. London: Royal Society of London, 1972.
3. Warburg OH. Über die katalytischen Wirkung der lebendigen Substanz, 1928; Schwerzmetalle als Wirkungen von Fermentaten, 1946 (Heavy Metals and Enzyme Action, 1949).

1932

☆

Physics
WERNER HEISENBERG
*for the creation of quantum mechanics, the application of which
led to the discovery of the allotropic forms of hydrogen.*

☆

Chemistry
IRVING LANGMUIR
for his discoveries and investigations in surface chemistry.

☆

Physiology or Medicine
SIR CHARLES SCHOTT SHERRINGTON and
LORD EDGAR DOUGLAS ADRIAN
for their discoveries regarding the functions of neurons.

☆

Literature
JOHN GALSWORTHY
*for his distinguished art of narration which takes
its highest form in The Forsythe Saga.*

☆

Peace
No Awards

1932

Edger Douglass Adrian (1889-1977) and Sir Charles Scott Sherrington (1857-1952)

Edgar Adrian   **Sir Charles Scott Sherrington**

© The Nobel Foundation

Charles Sherrington (right) **and Edger Adrian** were awarded the Nobel Prize in Physiology or Medicine "for their discoveries regarding the functions of neurons."

Sherrington was born in Islington, London. His father, a practicing doctor, died when Charles was a child and shortly thereafter his mother married Dr. Caleb Rose Jr., a noted archeologist and geologist. A man of many talents and wide-ranging interests, Dr. Rose exerted a profound influence on Charles and his brothers, William and George, both of whom became lawyers. Charles, however, became a scientist with a heart of a poet and soul of a philosopher.

Sherrington's scientific interests began when he was a boy, with the hobby of collecting and studying shells, fossils, and ancient coins. His poetic leanings and philosophical inclinations took shape in school, as he was being taught by the noted poet Thomas Ashe, and the distinguished scholar H. A. Holden. While he continued to foster his love for the written word, Sherrington also cultivated the habit of vigilantly exploring the world around him.

During his long, illustrious career, Sherrington elucidated the nature of the knee jerk reflex and advanced the basic concepts concerning cerebral localization and spinal cord functions. To describe the special nature of neuronal junctions, he coined the term "syndesm" (1887), but renamed it "synapse," as per the suggestions of Professor Verall, a friend at Trinity College and an Euripedean scholar. Sherrington also discovered and named proprioception, motor unit, stretch reflex, decerebrate rigidity, neuronal pool, reciprocal innervation, and many other concepts in neurophysiology. He was an expert skydiver, drawing huge crowds during parachute jumps from tops of London hospitals. Sir Charles had been called "William Harvey of Neurophysiology."

Lord Adrian, Chancellor of Cambridge University (1968-1975), was born in London and studied at Westminster and Cambridge. While a student under Keith Lucas at Cambridge, Adrian discovered that until one reached a certain level of intensity of electrical stimulation to the nerve, muscle contraction did not take place. Just beyond that threshold, however, all the muscle fibers contracted, a finding known as the "all-or-none phenomenon" of muscle contraction. After Lucas died in 1917, Adrian pursued a career in neurophysiology at Cambridge.

There he developed the capillary electrometer to measure minute signals from nerve fibers and discovered the nature of coding of the motor and sensory impulses in afferent and efferent fibers. In the early 1930s he advanced concepts of EEG, which had just been discovered by Hans Berger. Adrian once remarked that since his own brain produced excellent alpha waves, it was 'as good as a rabbit's brain!' He was a great teacher, expert fencer, and a legendary after-dinner speaker.

A man of tireless energy and industry, Adrian exerted a great influence on his pupils and on the development of neurophysiology research. The citizens of Cambridge looked upon him as an icon: his lean figure threading along the crowded, curvy roads of Cambridge on his bicycle at high speeds was an endearing sight.

Adrian was an enthusiastic mountaineer, a recreation he shared with Lady Adrian, who was a Justice of the Peace doing social work in Cambridge. Among Lord Adrian's other recreations were sailing and his great interest in the arts. Like his mentor and co-laureate, Lord Adrian was also deeply admired and respected throughout his life.

References

1. Adrian ED: The Impulses Produced by Sensory Nerve-Endings: Part IV. Impulses from Pain Receptors," Journal of Physiology, vol. 62, 1926; Adrian ED: "The Discharge of Impulses in Motor Nerve Fibres: Part I. Impulses in Single Fibres of the Phrenic Nerve," Journal of Physiology, vol. 66, 1928, p. 81 (with D.W. Bronk); "The Activity of Injury on Mammalian Nerve Fibres," Proceedings of the Royal Society; ser. B, vol. 106, p. 596; "The Activity of the Nervous System in the Caterpillar," Journal of Physiology, vol. 70, 1930.
2. Barlow HB. and Mollon JD, eds. The Senses. New York: Cambridge University Press, 1982.
3. Brazier, MAB. A History of Neurophysiology in the Nineteenth Century. New York: Raven Press, 1988.
4. Sherrington CS: Goethe on Nature and on science, 1942.
5. Sherrington CS: Journal of Physiology, vol. 5, 1884, p. vi
6. Sherrington CS: Man on His Nature, 1941
7. Sherrington CS: Notes on the Arrangement of Some Motor Fibres in the Lumbosacral Plexus. Journal of Physiology, vol. 13, 1892, pp. 621-772.

8. Sherrington CS: Notes on the Knee-Jerk and the Correlation of Action of Antagonistic Muscles. Proceedings of the Royal Society; vol. 52, 1893, pp. 556-564.

1933

☆

Physics
ERWIN SCHÖDINGER and PAUL ADRIEN MAURICE DIRAC
for the discovery of new productive forms of atomic theory.

☆

Chemistry
No Awards

☆

Physiology or Medicine
THOMAS HUNT MORGAN
for his discoveries concerning the role played by the chromosome in heredity.

☆

Literature
IVAN ALEKSEYEVICH BUNIN
*for the strict artistry with which he has carried on the
classical Russian traditions in prose writing.*

☆

Peace
SIR NORMALANGELL (RALPH LANE)
Writer. Member of the Executive Committee of the League
of Nations and the National Peace Council, and author
of the book *The Great Illusion*, among others.

1933

THOMAS HUNT MORGAN (1866-1945)

© The Nobel Foundation

Thomas Hunt Morgan was awarded the Nobel Prize in Physiology or Medicine "for his discoveries concerning the role played by the chromosomes in heredity."

He was born in Lexington, Kentucky, obtained Ph.D. from the Johns Hopkins University and joined Edward Wilson, a distinguished biologist at Columbia University in New York in 1904. Most of Morgan's early research came from this institution.

In 1900 a group of scientists had re-discovered Gregor Mendel's forgotten work on heredity and by 1904, chromosomes were the bearers of Mendelian "heredity factors." Although Morgan was skeptical of the chromosome theory, he began studying inheritance patterns in fruit flies, *Drosophila melanogaster*. The choice of this animal for genetic research turned out to be lucky for Morgan, since fruit flies were easy to obtain, they ate little and bred rapidly.

Morgan's fruit fly experiments revealed some contradictory findings to that of Mendel's laws of inheritance. One day in 1910 in his laboratory, fondly called the "fly room," Morgan found a male fly with white eyes rather than the usual red. Upon breeding this with a red-eyed female all first-generation offspring turned out to be red-eyed. Breeding those first-generation flies among themselves led to the second-generation offspring with both red and white-eyes in a ratio of 3:1. All white-eyed flies were males—Morgan had shown sex-linked inheritance. This work led to the discovery of the association between genes and chromosomes and the general role of chromosomes in transmitting genetic characters.

Morgan was an extremely modest scientist. He considered that the Nobel Prize was given in recognition of the new field of genetics rather than for of his research. Known for a great sense of humor, he often referred to his four children as "the F_1s," — the Mendelian notation for the first-generation.

References

1. Allen, Garland E. Thomas Hunt Morgan: The Man and His Science. Princeton, N.J.: Princeton University Press, 1978.
2. Carlson, Elof Alex. The Gene: A Critical History. Philadelphia: W.B. Saunders, 1966.
3. Morgan TH. Chromosomes and Heredity, American Naturalist, vol .44, 1910, pp. 449-496; Sex-Limited Inheritance in Drosophila, Science, vol. 32, 1910, pp. 120-122.
4. Morgan TH. Evolution and Adaptation, 1903; A Critique of the Theory of Evolution, 1916; Evolution and Genetics, 1925; What is Darwinism? 1929; The Scientific Basis of Evolution, 1932.
5. Morgan TH. The Development of the Frog's Egg, 1897; Experimental Embryology, 1927; Embryology and Genetics, 1934.

1934

Physics
No Awards

Chemistry
HAROLD CLAYTON UREY
for his discovery of heavy hydrogen.

Physiology or Medicine
GEORGE HOYT WHIPPLE, GOEORGE RICHARDS MINOT, and WILLIAM PARRY MURPHY
for their discoveries concerning liver therapy in cases of anemia.

Literature
LUIGI PIRANDELLO
for his bold and ingenious revival of dramatic and scenic art

Peace
ARTHUR HENDERSON
Former Foreign Secretary; Chairman of the League of Nations Disarmament Conference 1932-1934.

1934

GEORGE RICHARDS MINOT (1885-1950), WILLIAM PARRY MURPHY (1892-1987) AND GEORGE HOYT WHIPPLE (1878-1976)

George Minot   **William P. Murphy**   **George Whipple**
© The Nobel Foundation

George Minot (left), **William Murphy, and George Whipple** (right) shared the Nobel Prize in Physiology or Medicine "for their discoveries concerning liver therapy in cases of anemia."

Whipple was born in Ashland, New Hampshire, and took his degrees from Yale University and the Johns Hopkins Medical School, where he studied pathology under the famous William Welch. Here Whipple described a condition characterized by fat deposition in the intestinal lymphatics, later called Whipple's Disease.

From 1914 Whipple headed the Hooper Foundation in California. There he showed that liver, kidney and other meats stimulated hemoglobin production in the bone marrow, confirming Charles Hooper's suggestion that liver extracts might be useful in pernicious anemia. In 1936 Whipple was the first to suggest the term 'thalassemia' for the familial anemia seen among the people of Mediterranean descent.

Hematologist Minot was born in Boston and educated at Harvard Medical School. His research interests were in coagulation, functions of platelets and reticulocytes, leukemia and chronic anemia. Intrigued by Whipple's suggestion that pernicious anemia patients might lack a dietary substance, Minot used liver diets in a few patients who subsequently recovered. Minot then conducted a clinical trial at Boston's Peter Bent Bringham Hospital in collaboration with Murphy, a partner and a fellow Harvard graduate.

In 1926 Murphy and Minot reported that their 45 pernicious anemia patients, who had to be tube-fed with liver diets, were almost cured and became 'ravenously hungry' for liver. Further research confirmed the relationship between vitamin $B_{12,}$ liver and hemoglobin synthesis. It was shown that the "liver extract" contained vitamin B_{12}, which was critical for producing hemoglobin and curing anemia.

Initially, pernicious anemia patients had to eat a pound of liver daily. Manufacturing of liver extracts and vitamin B_{12} overcame such unpleasant requirement, saving millions of pernicious anemia patients the world over.

References

1. Castle WB. "A Century of Curiosity About Pernicious Anemia: The Gordon Wilson Lecture." Transactions of the American Clinical and Climatological Association 73 (1961): 54-80.
2. Corner GW. George Hoyt Whipple and His Friends: The Life of a Nobel Prize Pathologist. Philadelphia: J.B. Lippincott, 1963.
3. Minot GR. Origin of Antithrombin. American Journal of Physiology, vol. 38, 1915, pp. 233-247 (with George Denny)
4. Murphy WP. Treatment of Pernicious Anemia by a Special Diet. JAMA 87, 1926, pp. 470-476 (with George Minot).
5. Whipple GH. The Pathogenesis of Icterus. Journal of Experimental Medicine, vol. 13, 1911, pp. 115-135 (with John H. King);

Hametogenous and Obstructive Icterus: Experimental Studies by Means of the Eck Fistula. Journal of Experimental Medicine, vol. 17, 1913, pp. 593-611 (with Charles W. Hooper).

1935

☆

Physics
SIR JAMES CHADWICK
for the discovery of the neutron.

☆

Chemistry
FRÉDÉRIC JOLIOT and **IRÈNE JOLIOT-CURIE**
in recognition of their synthesis of new radioactive elements.

☆

Physiology or Medicine
HANS SPEMANN
for his discovery of the organizer effect in embryonic development.

☆

Literature
No Awards

☆

Peace
CARL VON OSSIETZKY
Journalist (with Die Weltbühne, among others), pacifist.

1935

HANS SPEMANN (1869-1941)

© The Nobel Foundation

Hans Spemann was awarded the Nobel Prize in Physiology or Medicine "for his discovery of the organizer effect in embryonic development."

He was born in Stuttgart, Germany. Following a year of military service and work at his father's book-selling business, Spemann returned to study Goethe, zoology, and physics. After graduating, he taught biology at the University of Würzburg and became Professor of Zoology at the University of Freiburg in 1919.

Spemann was one of the first experimental embryologists. He wished to learn the relationship between various segments of an embryo and subsequent body parts. Initially, he studied the salamander embryos by tying them off using thin suture-like hair. He tried to find whether one half of the embryo alone would lead to the formation of the complete organism. Subsequently, refining his already superb microsurgical techniques, Spemann identified specific regions in the embryos that had a capacity to 'induce' formation of new parts, when grafted onto other embryos. These regions he termed, the "organizer centers."

In subsequent works, Spemann and his colleagues showed that different organizer centers-controlled morphogenesis of the embryo overall and in its parts, such as the anterior regions into head and the posterior, into tail. They also discovered that the organizer function was not always predictable: when the posterior regions of the embryo were transplanted into the anterior regions, the recipient embryo formed two heads—not one head and a tail.

Because of his age, Spemann was spared from military service during the First World War and was allowed to continue research at Berlin's Kaiser Wilhelm Institute for Biology. It seems that when he learned that he had won Nobel Prize, Spemann exclaimed in characteristic modesty: "What can I do with so much money?"

References

1. Eakin, RM. Great Scientists Speak Again. Berkeley: University of California Press, 1975.
2. Fulton C, and Klein AO. Explorations in Developmental Biology. Cambridge, Mass.: Harvard University Press, 1976.
3. Moore R. The Coil of Life: The Story of the Great Discoveries in the Life Sciences. New York: Alfred A. Knopf, 1961.
4. Spemann H. "Experimentelle Beitrage zu einer Theorie der Entwicklung, 1936 (Embryonic Development and Induction, 1938); Forschung und Leben, 1943.

1936

☆

Physics
VICTOR FRANZ HESS
for his discovery of cosmic radiation, and
CARL DAVID ANDERSON
for his discovery of the positron.

☆

Chemistry
PETRUS (PETER) JOSEPHUS WILHELMUS DEBYE
*for his contributions to our knowledge of molecular structure
through his investigations on dipole moments and on
the diffraction of X-rays and electrons in gases*

☆

Physiology or Medicine
SIR HENRY HALLETT DALE and OTTO LOEWI
for their discoveries relating to chemical transmission of nerve impulses.

☆

Literature
EUGENE GLADSTONE O'NEILL
*for the power, honesty and deep-felt emotions of his dramatic
works, which embody an original concept of tragedy*

☆

Peace
CARLOS SAAVEDRA LAMAS
Foreign Minister. President of the League of Nations, Mediator
in a conflict between Paraguay and Bolivia in 1935.

1936

SIR HENRY HALLETT DALE (1875-1968)
AND OTTO LOEWI (1873-1961)

Henry Hallett Dale   **Otto Loewi**
© The Nobel Foundation

Sir Henry Hallett Dale (left) and **Otto Loewi** were awarded the 1936 Nobel Prize in Physiology or Medicine "for discoveries related to chemical transmission of nerve impulse."

Son of a businessman, Henry Dale was born in London, where he had early schooling. He graduated in Natural Sciences Tripos specializing in physiology and zoology from Trinity College in Cambridge. After a period of clinical training in St. Bartholomew Hospital of London, he received his medical degree from Cambridge University.

Dale studied under many leading physiologists of his day, including J. N. Langley, H.K. Anderson in Cambridge, and E.H. Starling in the University of London. While in London, Dale also met for the first time, young Otto Loewi with whom Dale would form a lifelong friendship. Dale also spent four valuable months carrying out research under the renowned scientist Paul Ehrlich in Frankfurt on Main in Germany. In 1904, he received his first job offer as pharmacologist from the Wellcome Physiological Research Laboratories in London. Despite the advice of his friends not to accept it, Dale took the job from "the world of commercial science," because he was assured that his research career would not be interfered with, and because he could earn a salary and get married.

Dale's early research was on the biochemistry of ergot—a chemical known to be present in fungus-contaminated rye bread, eating of which caused serious inflammation of the skin. However, because of its powerful effect on the smooth muscles of the uterus, extracts of ergot were also being used in obstetric practice during childbirth. Dale and colleagues extracted ergotoxine, an alkaloid, from ergot and demonstrated its vasodilatory action, an effect opposite to that of adrenaline. They extracted another powerful chemical from ergot, histamine, which had been artificially synthesized in 1907, but its presence was unknown in any natural substance. They also explained many of the major systemic effects of histamine.

The third chemical Dale extracted proved to be of even greater physiological significance: this was a product of reaction between acetic acid and choline, and it caused enlargement of blood vessels and slowed heartbeat—Dale christened it "acetylcholine." In a seminal paper in 1914, Dale detailed an account of the physiological effects of acetylcholine and noted that they were like those that followed the stimulation of parasympathetic nerves. He postulated that acetylcholine was probably 'the most suitable chemical for parasympathetic neurotransmission'— the first suggestion that chemicals may be involved in nerve impulse transmission.

Dale coined such key terms as 'cholinergic' and 'adrenergic' substances to describe chemicals mimicking acetylcholine and adrenaline and theorized that living tissues contained an enzyme that prevented the accumulation of acetylcholine. The enzyme was later isolated and named as *cholinesterase*.

Loewi was born in Frankfurt am Main. Persuaded by his parents, he chose to study medicine at Strasbourg, instead of classics and fine arts.

Having been acquainted with Henry Dale, F. G. Hopkins, E.H. Starling and other British scientists, Loewi pursued research in autonomous nervous system.

It seems that the classic experiment for which he is most famous occurred to him in a series of dreams during the Easter weekend of 1921. He then decided to try the same setup in his laboratory. The experiment consisted of using isolated hearts from two frogs. In one, the 'donor heart,' the vagus nerve was kept intact, while in the other, the 'recipient heart' the vagus was stripped. The hearts were connected using an inverted U-shaped glass tube, such that the saline solution perfusing the donor heart also perfused the recipient heart in a different chamber.

In this experimental setting, Loewi stimulated the vagus nerve of the donor heart, which led to the predictable slowing of the heartbeat, followed by its cessation. However, within 15 seconds of stimulation of the vagus, even the recipient heart began to slow down, eventually coming to a standstill. Loewi concluded that a chemical released from the donor heart after vagal stimulation was responsible for bradycardia in the recipient heart, just as it was responsible for a similar action in the donor heart. He named the chemical *"vagusstoff"* or the 'vagus substance,' in contrast to the *"acceleranstoff,"* or the accelerating substance (adrenaline). Later, Loewi proved that *vagusstoff* behaved like acetylcholine discovered by Dale and therefore must be one and the same.

Loewi and family endured horrible consequences of anti-Semitism in Europe. They were arrested but released only after surrendering all the Nobel Prize money to the Nazis. Initially, the Loewis Loewi emigrated to England and later to the United States of America.

References

1. Dale HH What Nagasaki Means, The Spectator, July 13, 1951.
2. Dale HH. The Action of the Pituitary Body, Biochemical Journal, vol. 4, 1909, p. 427

3. Dale HH. (with P. Laidlaw and C. Symons); Chemical Structure and Sympathomimetic Action of Amines, Journal of Physiology, vol. 41, 1910-1911, p. 19

4. Dale HH. A Reversed Action of the Vagus on the Mammalian Heart, Journal of Physiology, vol. 41, 1910, p. 318

5. Dale HH. On Some Physical Actions of Ergot, Journal of Physiology, vol. 34, 1906, p. 163

6. Dale HH. Science in War and Peace. Nature, vol. 160, 1947, p. 283.

7. Dale HH. The Atomic Problem Now. The Spectator, September 30, 1949

8. Dale HH: What I Believe, 1953; An Autumn Gleaning, 1954.

9. Feldberg, W.S. Sir Henry Hallett Dale. In Biographical Memoirs of Fellows of the Royal Society; Vol. 4. London: The Royal Society, 1970.

1937

☆

Physics
CLINTON JOSEPH DAVISSON and
SIR GEORGE PAGET THOMSON
for their experimental discovery of the diffraction of electrons by crystals.

☆

Chemistry
SIR WALTER NORMAN HAWORTH
for his investigations on carbohydrates and vitamin C, and
PAUL KARRER
for his investigations on carotenoids, flavins and vitamins A and B2.

☆

Physiology or Medicine
ALBERT von SZENT-GYÖRGYI NAGYRAPOLT
for his discoveries in connection with the biological combustion processes, with special reference to vitamin C and the catalysis of fumaric acid.

☆

Literature
ROGER MARTIN DU GARD
for the artistic power and truth with which he has depicted human conflict as well as some fundamental aspects of contemporary life in his novel cycle Les Thibault.

☆

Peace
CECIL OF CHELWOOD, VISCOUNT (LORD EDGAR ALGERNON ROBERT GASCOYNE CECIL)
Writer, Former Lord Privy Seal. Founder and President of the International Peace Campaign.

1937

ALBERT VON SZENT-GYÖRGYI (1893-1986)

© The Nobel Foundation

Albert Szent-Györgyi was awarded the Nobel Prize in Physiology or Medicine "for his discoveries related to biological combustion processes...vitamin C, and fumaric acid."

Szent-Györgyi was born into a distinguished family in Budapest. His medical studies at the University of Budapest were interrupted by the First World War, in which he served and was wounded. The political turmoil in Europe forced him to wander between Prague, Berlin, Hamburg, Leiden and Gröningen (in the Netherlands), Cambridge and other places. After the Soviet occupation of his country following World War II, he immigrated to the USA, where he spent the final decades of his life.

Szent-Györgyi chose biochemistry, since it is concerned with 'matters of life,' and began studying oxidative metabolism. Surprised at the similarity between the skin pigmentation in Addison's disease and that of the cut surfaces of potatoes exposed to air, he looked for biochemical explanations for these phenomena. After ten years of research, he postulated that a lack of an inhibitory, antioxidant substance found in oranges, lemons, and cabbages was responsible for color changes in potatoes. Because of its six-carbon structure he felt it to be a carbohydrate; this was named "hexuronic acid." Later Szent-Györgyi showed it to be vitamin C, found in large quantities in paprika. Walter Haworth of England discovered the structure of vitamin C, and together with Szent-Györgyi, named it "ascorbic acid" (from the Greek, "the anti-scurvy" factor).

About the time that Szent-Györgyi worked on ascorbic acid, Charles King at the University of Pittsburgh, working on the same topic published (before Szent-Györgyi did) an observation that certain crystals derived from lemon juice combated scurvy, and were chemically like hexuronic acid. As supporters of King claimed that he deserved to share the Nobel Prize, defenders of Szent-Györgyi countered that King had secretly known of Szent-Györgyi's results, which simply hastened his efforts at an earlier publication. These disputes stirred further controversy about the priority of the discovery of vitamin C, particularly after the announcement of the Nobel Prize for Szent-Györgyi.

Szent-Györgyi's war experiences profoundly influenced his moral and ethical stance. A lifelong pacifist, he undertook risky diplomatic missions and was haunted by the Nazis. In his book *The Crazy Ape* (1970) Szent-Györgyi questioned as to why humans behaved so irrationally, choosing war over peace, and death over life. He once wrote that he tried "to see what everyone else had seen, but to think what no one else had thought."

References

1. Fruton, Joseph S. Molecules and Life: Historical Essays on the Interplay of Chemistry and Biology. New York: Wiley-Interscience, 1972.
2. Moss RW. Free Radical Albert Szent-Gyorgyi and the Battle over Vitamin C. New York. Paragon House Publishers. 1988.

3. Szent-Gyorgyi A. Chemical Physiology of Contraction in Body and Heart Muscle, 1953.
4. Szent-Gyorgyi A. Science, Ethics, and Politics, 1963
5. Szent-Gyorgyi A. The Crazy Ape: Written by a Biologist for the Young, 1970
6. Szent-Gyorgyi A. What Next?! 1971.

1938

☆

Physics
ENRICO FERMI
*for his demonstrations of the existence of new radioactive elements
produced by neutron irradiation, and for his related discovery
of nuclear reactions brought about by slow neutrons.*

☆

Chemistry
RICHARD KUHN
*for his work on carotenoids and vitamins. (Caused by the authorities of his
country to decline the award but later received the diploma and the medal.)*

☆

Physiology or Medicine
CORNEILLE JEAN FRANÇOIS HEYMANS
*for the discovery of the role played by the sinus and aortic
mechanisms in the regulation of respiration.*

☆

Literature
PEARL BUCK (penname of PEARL WALSH née SYDENSTRICKER)
*for her rich and truly epic descriptions of peasant life in
China and for her biographical masterpieces.*

☆

Peace
NANSEN INTERNATIONAL OFFICE FOR REFUGEES
an international relief organization in Geneva
started by Fridtj of Nansen in 1921.

1938

CORNEILLE JEAN FRANÇOIS HEYMANS (1892-1968).

© The Nobel Foundation

Corneille Jean François Heymans was awarded the Nobel Prize in Physiology or Medicine "for the discovery of the role played by the sinus and aortic mechanisms in the regulation of respiration."

He was born in Ghent, Belgium, and received medical education at the University there, in which his father, Jan-Frans Heymans, Professor of Pharmacology, was a reputed scientist. After additional training in Paris, Vienna London and Cleveland, Ohio, Heymans joined the University of Ghent as lecturer. In 1930 he succeeded his father as the head of the Department of Pharmacology.

In the early 1920 it had been known that changes in blood pressure were associated with changes in respiration and heart rate. Some scientists had postulated that a few centers in the brain's medullary regions might be responsible for controlling these phenomena.

In 1924, Jean Heymans and his father began experiments in dogs. They had developed a very complex animal model. In one dog (referred to as dog B), they isolated most of the head from the rest of the body, which was kept alive by circulation provided from the trunk of another dog they referred to as dog A. This set up allowed them to independently change blood pressure of carbon dioxide concentrations in one animal, while studying the other, and vice-versa. They discovered when they stimulated the perfused head of dog B, the respirations increased solely in dog A. This indicated that the respiratory center was in the medulla of dog B and responded to the stimulation.

They also discovered that when they reduced the systemic blood pressure in dog A, the breathing in dog B became heavy. They thus established that the vagus nerve was the channel for transferring information from the brain to the distal organs, such as to breathing apparatus and blood pressure-controlling peripheral centers. These studies led to the discovery of chemoreceptors located at the cardio-aortic and carotid sinus regions.

Corneille Heymans also discovered the role of these receptors in proprioceptive regulation of arterial blood pressure and hypertension. He continued studies on cerebral circulation, the role of the kidney and sympathetic system in the control of blood pressure and other related works on pharmacology of the sympathetic nervous system. A prolific writer, Heymans authored more than 800 papers since 1920.

Heymans was a keen hunter and loved painting. He also maintained a deep interest in ancient literature and the history of medicine.

References

1. De Schaepdryver, A. F., ed. Corneel Heymans: A Collective Biography, Ghent, Belgium: Heymans Foundation, 1972.
2. Graubard, Mark. Circulation and Respiration: The Evolution of an Idea. New York: Harcourt, Brace & World, 1964.

3. Heymans CJF. The Control of Heart Rate Consequent to Changes in the Cephalic Blood Pressure and in the Intracranial Pressure, American Journal of Physiology, vol. 85, 1928. Pp. 498-506.

1939

☆

Physics
ERNEST ORLANDO LAWRENCE
for the invention and development of the cyclotron and for results obtained with it, especially with regard to artificial radioactive elements.

☆

Chemistry
ADOLF FRIEDRICH JOHANN BUTENANDT
for his work on sex hormones. (Caused by the authorities of his country to decline the award but later received the diploma and the medal), and
LEOPOLD STEPHEN RUZICKA
for his work on polymethylenes and higher terpenes.

☆

Physiology or Medicine
GERHARD DOMAGK
for the discovery of the antibacterial effects of Prontosil. (Was forced to decline the award, but after WW II, received the diploma and the medal.)

☆

Literature
FRANS EEMIL SILLANPÄÄ
for his deep understanding of his country's peasantry and the exquisite art with which he has portrayed their way of life and their relationship with Nature.

☆

Peace
No Awards

1939

GERHARD DOMAGK (1895-1964)

© The Nobel Foundation

Gerhard Domagk was awarded the Nobel Prize in Physiology or Medicine "for his discovery of the antibacterial effects of Prontosil."

Born in Lagow (now in Poland), Domagk studied medicine at Kiel University. He worked at University of Griefswald and the Pathological Institute at Münster prior to becoming the Director of Research at I.G. Farbenindustrie, a prominent dye-manufacturing company.

Following Paul Ehrilch's work, it had become possible to treat some protozoan and parasitic infections effectively, but bacterial infections continued to cause devastation. Domagk tested the antibacterial properties of the industrial dyes produced in his laboratory. In 1932 he found that a new azo dye, later named Prontosil, was remarkably effective against streptococcal sepsis when tested only *in vivo,* not *in vitro* in laboratory mice. He also felt that the drug was relatively safe. Domagk's daughter became one of the first patients to receive the new drug. She developed a serious streptococcal infection following a needle prick, and routine therapy was ineffective. She recovered upon receiving a large dose of Prontosil.

The news of Prontosil's effectiveness appeared in the German medical press even before Domagk's 1935 scientific publication on this subject, prompting clinical trials as early as 1933. The British Medical Research Council also confirmed Domagk's findings, and scientists from the Pasteur Institute discovered that the sulfanilamide portion of Prontosil was responsible for its curative properties. The sulfa drugs revolutionized treatment of bacterial infections, saving tens of thousands within a few years of their discovery.

Hitler forbade Germans from receiving Nobel Prizes and forced Domagk to decline it. But, having outlived the Nazi regime, Domagk received the Nobel diploma in 1947; however, there was no prize money, because as per the Nobel Foundation's statutes, it had been absorbed into the general pool.

References

1. Domagk G. Experimentelle Grundlagen der Chemotherapie mit Sulfonamiden, Antibiotica et Chemotherapia, vol. 4, 1957, pp. 1-45.
2. Schnitizer, R.J. and Frank Hawking, eds. Experimental Chemotherapy. Vol. 1 New York: Academic Press, 1963.

1940—1942 WWII YEARS

☆

NO AWARDS IN ANY CATEGORY

☆

1943

☆

Physics
OTTO STERN
for his contribution to the development of the molecular ray method and his discovery of the magnetic moment of the proton.

☆

Chemistry
GEORGE DE HEVESY
for his work on the use of isotopes as tracers in the study of chemical processes.

☆

Physiology or Medicine
HENRIK CARL PETER DAM
for his discovery of vitamin K, and
EDWARD ADELBERT DOISY
for his discovery of the chemical nature of vitamin K.

☆

Literature
No Awards

☆

Peace
No Awards

1943

HENRIK (CARL PETER) DAM
(1895-1976) AND EDWARD
ADELBERT DOISY (1893-1986)

Henrik Dam   **Edward Adelbert Doisy**
© The Nobel Foundation

Because of World War II, the Nobel Prizes in Physiology or Medicine were not awarded between 1940—42; the awards for 1943 and 1944 were presented together in New York in December 1944. **Henrik Dam** (left) and **Edward Doisy** shared the 1943 award for their independent work on vitamin K.

Dam was born in Copenhagen and studied at the Polytechnic Institute there. In 1929 he observed subcutaneous bleeding and delayed blood clotting in laboratory chicks raised on fat-free diets. By selective manipulation of diets, Dam first ruled out all known nutritional deficiencies, and later postulated the existence of an unknown chemical, since hog liver and some vegetable oils restored the clotting defect. He termed the substance the "K vitamin;" 'K' being the next letter available for assignment for a new vitamin. Coincidentally, it was also the first letter of "koagulation," or clotting in many Germanic languages. Later Dam worked with Paul Karrer of Zurich and extracted vitamin K from green vegetables and proved its effect on clotting.

Born in Hume, Illinois, Doisy studied at the University of Illinois and Harvard University and worked at St. Louis University. In 1923 Doisy and colleagues devised a bioassay for sex hormones and isolated estrion, and later estradiol. Turning to vitamin K research, in 1939, they successfully extracted vitamin K_1 from alfalfa and K_2 from fermented fish meal. Within a year the structure of vitamin K and its two forms had been elucidated, aiding its commercial production. Vitamin K soon became invaluable for treating liver diseases and preventing the hemorrhagic disease of the newborn.

Doisy always recognized his associates' contributions. He was so modest that he did not even refer to his Nobel Prize in his autobiography. A great benefactor to St. Louis University, Doisy donated to the university, all the profits from patents and sales of estrion, estradiol and vitamin K.

References

1. Dam H: Haemorrhages in Chicks Reared on Artificial Diets: A New Deficiency Disease, Nature, vol. 133 (June 16, 1934), pp. 909-910
2. Dam H: The Antihaemorrhage Vitamin of the Chick, Nature, vol. 135 (April 27, 1935), pp. 652-653.
3. Dam H: Vitamin K, Its Chemistry and Physiology, Advances in Enzymology, vol. 2, 1942, pp. 285-324.
4. Dam H: Vitamin K in Human Pathology, Lancet, March 26, 1938, pp. 720-721.
5. Doisy EA. An Ovarian Hormone, Journal of the American Medical Association, vol. 81, 1923, pp. 819-821 (with Edgar Allen)
6. Doisy EA: The Extract and Some Properties of an Ovarian

Hormone, Journal of Biological Chemistry, vol. 61, 1924, pp. 711-727 (with J.O. Ralls, Edgar Allen, and C.G. Johnston).

7. Doisy EA., The Crystals of the Follicular Ovarian Hormone, Proceedings of the Society for Experimental Biology and Medicine, vol. 27, 1930, pp. 417-419 (with Sidney Thayer and Clement Veler)

8. Fitch, Coy D. In Memoriam: Edward A. Doisy. Endocrinology 122 (1988): 2348-2349.

1944

☆

Physics
ISIDOR ISAAC RABI
for his resonance method for recording the magnetic properties of atomic nuclei.

☆

Chemistry
OTTO HAHN
for his discovery of the fission of heavy nuclei.

☆

Physiology or Medicine
JOSEPH ERLANGER, and HERBERT SPENCER GASSER
*F\for their discoveries relating to the highly
differentiated functions of single nerve fibers.*

☆

Literature
JOHANNES VILHELM JENSEN
*for the rare strength and fertility of his poetic imagination
with which is combined an intellectual curiosity of
wide scope and a bold, freshly creative style.*

☆

Peace
INTERNATIONAL COMMITTEE OF THE RED CROSS

1944

JOSEPH ERLANGER (1874-1965) AND HERBERT SPENCER GASSER (1888-1963)

Joseph Erlanger   **Herbert Spencer Gasser**
© The Nobel Foundation

Joseph Erlanger (left) and **Herbert Spencer Gasser** were jointly awarded the 1944 Nobel Prize in Physiology or Medicine "for their discoveries regarding the highly differentiated functions of single nerve fibers."

Born in San Francisco, Erlanger studied chemistry at the University of California, Berkeley, and medicine at the Johns Hopkins Medical School. In 1906, he became the first professor of physiology at the new medical school of the University of Wisconsin in Madison. In 1910 he moved to the Washington University in St. Louis.

Gasser was born in Platteville, Wisconsin and studied at University of Wisconsin's Madison campus, where he was Erlanger's student. After receiving his MD degree from the Johns Hopkins University in 1915, he rejoined Erlanger in St. Louis at Washington University.

It had long been known that nerve fibers transmit sensory and motor impulses in the form of electrical signals. However, these signals were so weak and brief, that they could not be measured accurately. Gasser and Erlanger developed equipment that could amplify the nerve impulses by several thousand-fold. They built oscilloscopes to display the impulses as waveforms on fluorescent screens.

Linking amplifiers to oscilloscopes and attaching sensors to individual frog nerve fibers, Gasser, Erlanger and coworkers showed that nerve conduction properties depended upon the fiber size and function; the thick motor nerve fibers conducted the fastest, whereas the thin, pain-carrying fibers, the slowest. Like Edgar Adrian's work, these findings profoundly influenced physiology research in the 20th century.

Because of the Second World War, the Nobel prizes for 1943 and 1944 were distributed in a special function on December 11, 1944, in New York—the only time such ceremonies to be held in the United States.

References

1. Bergland R. The Fabric Mind. Ringwood, Australia: Penguin Books Australia, 1985.
2. Erlanger J. A Study of the Metabolism in Dogs with Shortened Small Intestines, American Journal of Physiology, vol. 6, 1901, pp. 1-30 (with W. E)
3. Erlanger J. Studies in Blood Pressure Estimations by Indirect Methods: II. The Mechanism of the Compression Sound of Korotkoff, American Journal of Physiology, vol. 40, 1916, pp. 82-125
4. Erlanger J. The Compound Nature of the Action Current of Nerve as Disclosed by the Cathode-Ray Oscillograph, American Journal of Physiology, vol. 70, 1924, pp.624-666 (with H.S. Gasser)

5. Erlanger J. Electrical Signs of Nervous Activity, 1937 (with H.S. Gasser)

6. Erlanger, Joseph. A Physiologist Reminisces. In Annual Review of Physiology, edited by Victor E. Hall. Palo Alto, CA. Annual Reviews, 1964.

7. Gasser HS (with Newcomer HS). Physiological Action Currents in the Phrenic Nerve: An Application of the Thermionic Vacuum Tube to Nerve Physiology. American Journal of Physiology, vol. 57, pp. 1-26, 1921.

8. Gasser HS (with Erlanger J). A Study of the Action Currents of Nerve with the Cathode Ray Oscillograph. American Journal of Physiology, vol. 62, pp. 496-524, 1922.

1945

☆

Physics
WOLFGANG PAULI
for the discovery of the **Exclusion Principle,** *also called the* **Pauli Principle.**

☆

Chemistry
ARTTURI ILMARI VIRTANEN
*for his research and inventions in agricultural and nutrition
chemistry, especially for his fodder preservation method.*

☆

Physiology or Medicine
SIR ALEXANDER FLEMING, SIR ERNST BORIS
CHAIN and LORD HOWARD WALTER FLOREY
*for the discovery of penicillin and its curative
effect in various infectious diseases.*

☆

Literature
GABRIELA MISTRAL
*for her lyric poetry which, inspired by powerful emotions, has made her name
a symbol of the idealistic aspirations of the entire Latin American world.*

☆

Peace
CORDELL HULL
Former Secretary of State. One of the initiators of the United Nations.

1945

ERNST BORIS CHAIN (1906-79);
SIR ALEXANDER FLEMING (1881-
1955); AND BARON HOWARD
WALTER FLOREY (1898-1968)

Sir Ernst Boris Chain   Sir Alexander Fleming   Howard Walter Florey
© The Nobel Foundation

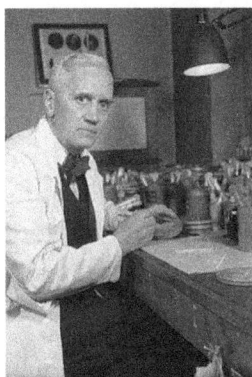

Alexander Fleming (right), **Ernst Boris Chain** (left), and **Baron Howard Walter Florey** shared the 1945 Nobel Prize in Physiology or Medicine "for the discovery of penicillin and its therapeutic effects."

An Ayrshire farmer's son from Scotland, Fleming studied medicine at Saint Mary's Hospital Medical School in London, where he worked under Sir Almroth Wright.

In 1921, Fleming discovered and named lysozyme from human nasal secretions; he thought that lysozyme might possess antiseptic properties. In 1928, he was studying the shapes and colors of bacteria for a chapter in an encyclopedic volume. While he was away on vacation, an accidental mold-contamination of one of his culture plates had inhibited staphylococcal growth. Instead of discarding the specimen, Fleming studied this mold, *Penicillum notatum*, and named its bactericidal chemical, "penicillin." He published this discovery in 1929, but the work remained obscure for 11 years, in part because Fleming's main thrust in the paper was on the biology of the mold.

Born in Adelaide, Australia, Florey studied medicine there. He worked in Cambridge and later in Oxford, where he collaborated with Chain—the Berlin-born scientist who had emigrated to Oxford in 1933 to escape Hitler's Germany. Florey and Chain were interested in developing antiseptic agents from natural substances, and thus they considered penicillin. After painstaking research, they isolated penicillin in pure form in 1939 and proved in clinical studies, its remarkable safety and efficacy. They developed steps for penicillin production and obtained help from the United States government and pharmaceutical companies for its manufacturing. By 1941, some penicillin became available for use among the allied soldiers during World War II.

The news media and the public hailed as legendary, the story of Fleming's accidental discovery of penicillin. Yet, as Göran Liljestrand of the Nobel Foundation noted (and Fleming concurred), without Florey and Chain, penicillin would have stayed "a fairly unknown substance, interesting to the bacteriologist, but of no practical importance."

References

1. Abraham EP. Howard Walter Florey. Biographical Memoirs of Fellows of the Royal Society 17, pp. 255-302, 1971.
2. Colebrook L. Alexander Fleming. Biographical Memoirs of the Royal Society 2, pp. 117-127, 1956.
3. Fleming SA. A Comparison of the Activities of Antiseptics on Bacteria and Leucocytes. Proceedings of the Royal Society, vol. B96, p. 171, 1924
4. Fleming SA. On the Antibacterial Actions of Cultures of a Penicillium, with Special Reference to Their Use in the Isolation of B. Influenzae. British Journal of Experimental Pathology, vol. 10, p. 226, 1929.

5. Fleming SA. Observations on a Bacteriolytic Substance (Lysozyme) Found in Secretions and Tissues. British Journal of Experimental Pathology, vol. 3, p. 252, 1922

6. Fleming SA. On a Remarkable Bacteriolytic Substance Found in Secretions and Tissues. Proceedings of the Royal Society, vol. B93, p. 306, 1922

7. Florey B (with Chain E, Gardner AD, Heatley NG, Jennings MA, Orr-Ewing J and Sanders AG). Penicillin as a Chemotherapeutic Agent. Lancet, pp. 226-233, August 24, 1940; (with Abraham EP, Chain E, Fletcher CM, Gardner AD) Further Observations on Penicillin. Lancet, pp. 177-185, August 16, 1941.

8. Macfarlane G. Alexander Fleming. The Man and the Myth. Cambridge, Mass.: Harvard University Press, 1984.

1946

☆

Physics
PERCY WILLIAMS BRIDGMAN

*for the invention of an apparatus to produce extremely high pressures, and
for the discoveries he made therewith in the field of high-pressure physics.*

☆

Chemistry
JAMES BATCHELLER SUMNER

for his discovery that enzymes can be crystallized.

JOHN HOWARD NORTHROP and
WENDELL MEREDITH STANLEY

for their preparation of enzymes and virus proteins in a pure form.

☆

Physiology or Medicine
HERMANN JOSEPH MULLER

for the discovery of the production of mutations by means of X-ray irradiation.

☆

Literature
HERMANN HESSE

*for his inspired writings which, while growing in boldness and penetration,
exemplify the classical humanitarian ideals and high qualities of style.*

☆

Peace
EMILY GREENE BALCH

Former Professor of History and Sociology, Honorary International
President Women's International League for Peace and Freedom, and

JOHN RALEIGH MOTT

Chairman of the first International Missionary Council, President
of the World Alliance of Young Men's Christian Associations.

1946

HERMANN JOSEPH MULLER (1890-1967)

© The Nobel Foundation

Hermann Muller was awarded the 1946 Nobel Prize in Physiology or Medicine "for the discovery of mutations by means of X-ray radiation."

Muller was born in New York City, studied and taught at New York's Columbia College and Cornell University. Thomas Hunt Morgan, the 1934 Nobel Laureate, was his mentor at Columbia. After three years at Rice Institute in Houston, Muller moved to the University of Texas in Austin in 1915, where he worked until 1932.

Muller carried out a series of studies testing the natural mutation rates in the fruit fly, *Drosophilae melanogaster*. He found that raising the environmental temperature increased the mutation rate. He also discovered that mutation rates increased several-fold upon exposures to X-rays. Muller's 1929 paper, "Artificial transmission of the gene," in *Science* became sensational, as he had shown that natural mutations were rare and inconsequential, whereas even diagnostic X-rays could induce dangerous mutations.

Muller became deeply concerned about needless exposures to radiation in diagnostic radiology and radiation therapy. In the 1930s, it was common to administer massive doses of irradiation to the ovaries to induce ovulation in infertile women. Paradoxically, irradiation of the testes was used to render men temporarily sterile. Muller pointed out the fallacy of these therapies and argued that hazards of radiation may manifest in offspring several generations later.

A severe critic of the nuclear race, Muller vehemently opposed the US policy of testing and using atomic weapons. Despite his highly controversial view of creating a utopian world through eugenics, Muller is remembered for his scientific contributions, his far-reaching humanitarian concerns, and his social activism.

References

1. Carlson EA. H.J. Muller. Genetics 70, pp. 1-30 (January 1972).
2. Joravsky D. The Lysenko Affair. Scientific American 207 pp. 41-49, November 1962.
3. Muller HJ (with Little CC, Snyder LH). Out of the Night: A Biologist's View of the Future. Genetics, Medicine, and Man, 1947.
4. Muller HJ (with Morgan TH, Sturtevant AH and Bridges CB). The Mechanism of Mendelian Heredity. Artificial Transmutation of the Gene. Science, vol. 66, pp. 84-87, July27, 1927.

1947

Physics
SIR EDWARD VICTOR APPLETON

for his investigations of the physics of the upper atmosphere especially for the discovery of the so-called Appleton layer.

☆

Chemistry
SIR ROBERT ROBINSON

for his investigations on plant products of biological importance, especially the alkaloids.

☆

Physiology or Medicine
CARL FERDINAND CORI, and GERTY THERESA CORI

for their discovery of the course of the catalytic conversion of glycogen; and
BERNARDO ALBERTO HOUSSAY

for his discovery of the part played by the hormone of the anterior pituitary lobe in the metabolism of sugar.

☆

Literature
ANDRÉ PAUL GUILLAUME GIDE

for his comprehensive and artistically significant writings, in which human problems and conditions have been presented with a fearless love of truth and keen psychological insight.

☆

Peace
THE FRIENDS SERVICE COUNCIL (The Quakers), London. Founded in 1647, and
THE AMERICAN FRIENDS SERVICE COMMITTEE (The Quakers), Washington.

The society's first official meeting was held in 1672.

1947

Carl Ferdinand Cori (1896-1984); Gerty Theresa Radnitz Cori (1896-1957); and Bernardo Alberto Houssay (1887-1971)

Carl Ferdinand Cori   Gerty Theresa Cori   Bernardo Alberto Houssay

© The Nobel Foundation

The 1947 Nobel Prize in Physiology or Medicine was jointly awarded to **Bernardo Houssay** (bottom) and the **Cori couple** for their independent works related to glucose metabolism.

Carl and Gerty Cori were both born in Prague and obtained their MD degrees from the German University of Prague in 1920—the year they were married. Political turmoil forced them to emigrate to the US in 1922, where they worked as biochemists first at the University of Buffalo, New York, and later at Washington University, St. Louis.

The Coris studied glucose metabolism. Using washed, rinsed, frog muscles, they discovered glucose-1-phosphate (the "Cori ester"), as intermediary agent between glycogen and glucose-6-phosphate. They also discovered phosphorylase, the enzyme involved in glycogen breakdown, and proposed in 1929, "a cycle of carbohydrates" (the Cori cycle), in which blood glucose becomes muscle glycogen, which breaks down to blood lactic acid, which in turn becomes liver glycogen. Although the Cori cycle has been modified, its essential components have remained unchanged.

Forty-six years had passed since its inception before a woman, Gerty Cori, was to win a Nobel Prize in Physiology or Medicine. For much of the 20th century, female scientists had to overcome numerous hurdles to make their mark in the elite male-dominated world of scientific research. In a job interview, Gerty Cori was told not 'to stand in the way of her husband Carl's career,' and her husband was told that it was "un-American" for him to work with his wife. Karl Cori was also told that he had an accent, and hence, ought not to consider any academic position. Thus, it is to their great credit that despite numerous setbacks, the Cori couple prevailed and excelled in their scientific pursuits.

Houssay was born in Buenos Aires and graduated in medicine in 1909. He soon became the professor and director of the Institute of Physiology there. Houssay and coworkers showed that the pituitary was the "master gland," modulating metabolic functions. They noted that the removal of the anterior lobe of the pituitary rendered tdogs insulin sensitive, while pituitary extracts made them insulin resistant and diabetic. Houssay correctly hypothesized that this was because of growth hormone.

As was the case with the Coris, political upheavals affected Houssay's career, too. Because of his vocal opposition to the Perón's regimen, Houssay was fired in 1943 but was re-appointed a decade later. In the meantime, he had founded the Institute of Biology and Experimental Medicine in Argentina, which became an internationally renowned research center.

References

1. Cori CF, The call of science. *Amer Rev Biochem* 1969;38;1-20.
2. Houssay BA. Fisiología humana, 1946 (Human Physiology, 1948.

3. Houssay BA. La acción diabetógena de la hipófisis, 1945

4. Houssay BA. Collected Papers on Medical Subjects Including Physiology, 1924-1934

5. Houssay BA. Functions of the Pituitary Gland, 1936

6. Houssay BA. In Current Biography 1948, edited by Anna Rothe. New York: H.W. Wilson, 1949.

1948

Physics
LORD PATRICK MAYNARD STUART BLACKETT
for his development of the Wilson cloud chamber method, and his discoveries therewith in the fields of nuclear physics and cosmic radiation.

☆

Chemistry
ARNE WILHELM KAURIN TISELIUS
for his research on electrophoresis and adsorption analysis, especially for his discoveries concerning the complex nature of the serum proteins.

☆

Physiology or Medicine
PAUL HERMANN MÜLLER
for his discovery of the high efficiency of DDT as a contact poison against several arthropods.

☆

Literature
THOMAS STEARNS ELIOT
for his outstanding, pioneer contribution to present-day poetry.

☆

Peace
No Award

1948

PAUL HERMANN MÜLLER (1899-1965)

© The Nobel Foundation

Paul Hermann Müller was awarded the 1948 Nobel Prize in Physiology or Medicine "for his development of DDT as an insecticide."

Born in Olten, Switzerland, Müller studied chemistry at the University of Basel. In 1925 he joined the J. R. Geigy pharmaceutical company where he initially did research on industrial dyes, and in 1935 switched to developing insecticides.

The existing insecticides were costly, lacked effectiveness, or were too toxic. Müller hoped to develop insecticides with long-lasting efficacy that would be effective against pests, but safe to other animals and plants. After testing hundreds of compounds, in 1939, Müller and colleagues discovered that *d*ichloro-*d*iphenyl-*t*richloroethane, or DDT, was a potent insecticide. DDT had been synthesized in Germany some 60 years earlier, yet its pesticidal properties had not been studied.

Soon after its commercial release in the 1940s, DDT became a phenomenal success worldwide. During World War II it helped prevent malaria by eradicating mosquitoes in the South Pacific and South China Sea regions where the allied soldiers were in action. Using DDT as skin powder, nearly 1.5 million people were deloused to halt an ongoing typhus epidemic in Naples in 1943-44. A similar approach-controlled typhus in Japan in 1945.

But, as early as in 1946, Müller became concerned about the long-term side effects of DDT. Instead of degrading into inert products, DDT accumulated unaltered in tissues. It killed useful insects by depriving them of their natural habitats. In the 1960s and 70s, environmental activists rallied against DDT, and the US and other countries banned it. Despite its eventual downfall, we must note that, as the most effective agricultural pesticide ever developed, DDT played an important role during a pivotal period in our history.

References

1. Asimov I. Paul Hermann Müller. In Asimov's Biographical Encyclopedia of Science and Technology. Garden City, N.Y.: Doubleday, 1982.
2. Campbell GA. "Dr. Paul Hermann Müler." *Chemistry and Industry* 52 (December 25, 1965) 2105-2106.
3. Dunlap TR. *DDT: Scientists,Citizens, and Public Policy*. Princeton, Princeton University Press.1981.
4. Zimmerman OT and Lavin I. *DDT: Killer of Killers*. Dover NH. Industrial Research Service, 1946.

1949

☆

Physics
HIDEKI YUKAWA
*for his prediction of the existence of mesons on the
basis of theoretical work on nuclear forces.*

☆

Chemistry
WILLIAM FRANCIS GIAUQUE
*for his contributions in the field of chemical thermodynamics, particularly
concerning the behavior of substances at extremely low temperatures.*

☆

Physiology or Medicine
WALTER RUDOLF HESS
*for his discovery of the functional organization of the interbrain
as a coordinator of the activities of the internal organs; and*
ANTONIO CAETANO DE ABREU FREIRE EGAS MONIZ
for his discovery of the therapeutic value of leucotomy in certain psychoses.

☆

Literature
WILLIAM FAULKNER
*for his powerful and artistically unique contribution
to the modern American novel.*

☆

Peace
LORD JOHN BOYD ORR OF BRECHIN
Physician, Alimentary Politician, prominent organizer and Director
General Food and Agricultural Organization, President National
Peace Council and World Union of Peace Organizations

1949

WALTER RUDOLF HESS (1881-1973);
ANTÔNIO EGAS MONIZ (1874-1955)

Walter Rudolf Hess   **António Egas Moniz**

© The Nobel Foundation

The 1949 Nobel Prize in Physiology or Medicine was shared between **Walter Rudolf Hess** (left) "for his work related to the diencephalon," and **Antônio Egas Moniz** "for his discovery of the therapeutic value of frontal leucotomy."

Hess was born in Frauenfeld, Switzerland. After graduating in medicine in 1906, he practiced ophthalmology for 11 years before accepting a research position at the University of Zürich's Physiology Institute. Initially, he worked on cardiovascular control mechanisms, and in 1925, switched to studying the autonomic nervous system. Using electrical stimulation in non-anesthetized cats, Hess discovered that the diencephalon, the hypothalamus in particular, was crucial for controlling blood pressure, respiration, temperature and hunger, and for modulating fear and anger. Hess summarized 25-years of his research in two classic monographs in 1947. He continued research after retirement and helped found an institute for brain research in Zürich.

Born in Avança, Portugal, Antônio Egas Moniz became head of neurology at the University of Lisbon in 1911. At a conference in London in 1935, Moniz learned from John Fulton's work, that bilateral frontal lobotomy rendered otherwise excitable monkeys, docile and timid. Postulating abnormal frontal lobe connections in psychiatric patients, Moniz conceived similar procedures in humans. In 1936, with help from colleague Almeida Lima, Moniz performed the first frontal 'leucotomy' (or lobotomy) in a psychotic patient. Initially he used alcohol injection to sever the white matter connections. Later he adopted surgical resection of portions of frontal lobes.

Although there were critics, Moniz's "psychosurgery" got accepted; over 5000 lobotomies were performed between 1942 and 1952 in the US alone. The advent of anti-psychotic drugs in the 1960s, however, rendered lobotomy obsolete. Medical historians rightly criticize Moniz for his unsubstantiated claims, disregard for complications and poor follow-up. Yet, Moniz is remembered as an eminent neurologist who introduced cerebral angiography—perhaps his legacy. He also served Portugal as a parliamentarian, its ambassador in Spain, and its foreign minister.

References

1. Glaser H. The Road to Modern Surgery: The Advances in Medicine and Surgery During the Past Hundred Years. Translated by Maurice Michael. London: Lutterworth Press, 1960.
2. Koskoff YD, Goldhurst R. The Dark Side of the House. New York: Dial Press, 1968.
3. Moniz AE. Prefrontal Leucotomy in the Treatment of Mental Disorders. American Journal of Psychiatry, vol. 93, pp. 1379-1385, 1937.

4. Valenstein ES. Great and Desperate Cures: The Rise and Decline of Psychosurgery and Other Radical Treatments for Mental Illness. New York: Basic Books, 1986.

1950

☆

Physics
CECIL FRANK POWELL
for his development of the photographic method of studying nuclear processes and his discoveries regarding mesons made with this method.

☆

Chemistry
OTTO PAUL HERMANN DIELS and KURT ALDER
for their discovery and development of the diene synthesis.

☆

Physiology or Medicine
EDWARD CALVIN KENDALL, TADEUS REICHSTEIN and PHILIP SHOWALTER HENCH
for their discoveries relating to the hormones of the adrenal cortex, their structure and biological effects.

Literature

☆

EARL BERTRAND ARTHUR WILLIAM RUSSELL
in recognition of his varied and significant writings in which he champions humanitarian ideals and freedom of thought.

☆

Peace
RALPH BUNCHE
Professor Harvard University, Cambridge, MA; Director of the UN Division of Trusteeship, Acting Mediator in Palestine 1948.

1950

EDWARD CALVIN KENDALL (1886-1972)
PHILIP SHOWALTER HENCH (1896-1965)
AND ADEUS REICHSTEIN (1897-1996)

Philip Showalter Hench   Edward Calvin Kendall   Tadeusz Reichstein

© The Nobel Foundation

Edward Kendall (right), **Philip Hench** (left), and **Adeus Reichstein** shared the Nobel Prize in Physiology or Medicine in 1950 "for discoveries concerning suprarenal cortex hormones."

Born in South Norwalk, Connecticut, Kendall graduated from New York's Columbia University and joined the Mayo Clinic in 1914, where he pursued research on thyroid hormones. On December 23, 1914, he became the first to isolate thyroxin in crystalline form enabling its large-scale manufacturing.

Hench was born and educated in Pittsburgh and obtained clinical training at the Mayo Clinic; there in 1926 he headed the new rheumatic disease section. Some of his arthritis patients used to feel their joint symptoms improved if they developed jaundice, and others when they became pregnant. To explain these phenomena, Hench postulated that an anti-rheumatic chemical, or "substance X," is produced in the body in response to stress. In 1935 he began collaborating with Kendall to identify substance X.

The Park Davis Company had been supplying Kendall bovine adrenal glands at no cost, in return, he was to send them the extracted adrenaline. Using the left-over cortex (from the supplied adrenal glands) he had isolated four new adrenal cortical hormones by 1935, which he named "compounds A, B, E, and F." Hench, Kendall and associates later found that Kendall's compound E and Hench's substance X was one and the same. Eight years later they extracted small quantities of compound E for clinical trials. Hench named it 'cortisone.'

Tadeus Reichstein was born in Wloclawek, near Warsaw. His family moved to Zurich and took up Swiss citizenship in 1914. A tireless worker, Reichstein's initial research was on the chemistry of coffee and ascorbic acid. Switching to research on hormone biochemistry, he discovered 27 chemicals from the adrenal glands and deciphered the structure of many, including that of aldosterone. Reichstein perfected techniques for manufacturing adrenal hormones heralding a new era in therapeutic pharmacology.

References

1. Hench PS, Kendall EC, Slocumb CH and Polley HF. Effects of Cortisone Acetate and Pituitary ACTH on Rheumatoid Arthritis, Rheumatic Fever, and Certain Other Conditions. Archives of Internal Medicine 85, pp. 545-666, 1950.
2. Hench PS. Effect of Jaundice on Chronic Infectious (Atrophic) Arthritis and on Primary Fibrositis. Archives of Internal Medicine, vol. 61, pp. 451-480, 1938.
3. Hench PS. The Present Status of Cortisone and ACTH in General Medicine. Proceedings of the Royal Society of Medicine 43, pp. 769-773, 1950.
4. Kendall EC (with Hench PS, Slocumb CH, Polley HF). The Effect of the Hormone of the Adrenal Cortex (17-Hydroxy-

11-Dehydrocorticosterone: Compound E) and of Pituitary Adrenocorticotropic Mayo Clinic, vol 24, 1949.

5. Kendall EC. Studies Related to the Adrenal Cortex. Federation Proceedings, vol. 9, 1950.
6. Kendall EC. The Hormone of the Adrenal Cortex Designated Compound E. Proceedings of the Staff Meetings of the Mayo Clinic, vol. 24, 1949
7. Kendall EC. The Influence of the Adrenal Cortex on the Metabolism of Water and Electrolytes. Vitamins and Hormones, vol. 6, 1948
8. Kendall EC. Thyroxin, American Chemical Society Monograph 47, 1928
9. Oesper R. Tadeus Reichstein. Journal of Chemical Education 26, pp. 529-530, October 1949.
10. Reichstein T (with Shoppee CW). The Hormones of the Adrenal Cortex, in Vitamins and Hormones, edited by R.S. Harris and K.V. Thiamin, 1943

1951

☆

Physics
SIR JOHN DOUGLAS COCKCROFT and
ERNEST THOMAS SINTON WALTON
*for their pioneer work on the transmutation of atomic
nuclei by artificially accelerated atomic particles.*

☆

Chemistry
EDWIN MATTISON MC MILLAN, and
GLENN THEODORE SEABORG
for their discoveries in the chemistry of the transuranium elements.

☆

Physiology or Medicine
MAX THEILER
for his discoveries concerning yellow fever and how to combat it.

☆

Literature
PÄR FABIAN LAGERKVIST
*for the artistic vigor and true independence of mind with which he endeavors
in his poetry to find answers to the eternal questions confronting mankind.*

☆

Peace
LÉON JOUHAUX
President of the International Committee of the European Council, Vice
President of the World Federation of Trade Unions; delegate to the UN.

1951

MAX THEILER (1899-1972)

© The Nobel Foundation

Max Theiler was awarded the 1951 Nobel Prize in Physiology or Medicine "for his discoveries concerning yellow fever and how to contain it."

He was born in Pretoria and studied medicine at London's Saint Thomas Hospital. In 1922 he moved to the United States and received training at Harvard's Tropical Medicine department and later worked there. In 1930 he joined the Rockefeller Foundation to pursue research on infectious diseases.

By the early 1920s it had been known that yellow fever was caused by a filterable virus and transmitted through the mosquito *Aedes aegypti*. Yellow fever research was clearly very dangerous—five scientists, including the famous bacteriologist Hideyo Noguchi, had died of it studying this highly contagious disease.

Using frozen liver tissues from an experimentally infected monkey, Theiler induced yellow fever encephalomyelitis in mice and discovered that strains re-isolated from them were less virulent than the original strains. It was known that survivors of yellow fever acquired a permanent immunity. Theiler injected the laboratory mice with combinations of yellow fever virus and serum from volunteers with a history of yellow fever. In such mice he showed that yellow fever immunity can be induced.

By the mid-1930s, Theiler had become famous for developing the mouse model of yellow fever; using this he developed a vaccine (for use in humans) from the attenuated strain of the virus. However, the vaccine was too toxic, causing encephalitis in some recipients. In 1937 Theiler developed a further attenuated, "17 D vaccine," using chick embryo cultures.

Between 1940 and 1947, 28 million people in Africa and the Americas were inoculated with this vaccine, protecting them from the ravages of yellow fever. In later years, Theiler described an encephalomyelitis in mice identical to human polio; this is sometimes called *Theiler's disease*.

References

1. Strode GK. Ed. Yellow Fever. New York: McGraw-Hill, 1951.
2. Theiler M. Studies on the Action of Yellow Fever Virus in Mice. Annals of Tropical Medicine and Parasitology, vol. 24, pp. 249-272, 1930
3. Theiler M. (with Kitchen SF, Lloyd WDM) Vaccination Against Yellow Fever with Immune Serum and Virus Fixed for Mice. Journal of Experimental Medicine, vol. 55, p. 945-969, 1932.
4. Theiler M. Susceptibility of White Mice to Virus of Yellow Fever. Science, vol. 71, p. 367, 1930

1952

☆

Physics
FELIX BLOCH and **EDWARD MILLS PURCELL**
for their development of new methods for nuclear magnetic precision measurements and discoveries in connection therewith.

☆

Chemistry
ARCHER JOHN PORTER MARTIN and **RICHARD LAURENCE MILLINGTON SYNGE**
for their invention of partition chromatography.

☆

Physiology or Medicine
SELMAN ABRAHAM WAKSMAN
for his discovery of streptomycin, the first antibiotic effective against tuberculosis.

☆

Literature
FRANÇOIS MAURIAC
for the deep spiritual insight and the artistic intensity with which he has in his novels penetrated the drama of human life.

☆

Peace
ALBERT SCHWEITZER
Missionary surgeon, Founder Lambaréné
Hospital in République du Gabon.

1952

SELMAN ABRAHAM
WAKSMAN (1888-1973)

© The Nobel Foundation

Selman Abraham Waksman received the Nobel Prize in Physiology or Medicine "for his discovery of streptomycin."

Born in Priluka near Kiev, Waksman immigrated to the United States in 1910 and studied soil bacteriology at the Agriculture College of Rutgers University. After years of tedious research using soil samples from the campus at Rutgers, Waksman discovered and characterized numerous soil microbial florae. He proved that humus and peat resulted from degradation of plant and animal waste products.

By 1939, René Dubos, Waksman's postdoctoral student, had extracted from *Bacillus brevis*, 'tyrocidine' and 'gramicidin,' two germicides effective against bacterial infections in the cattle, but too toxic for humans. Pursuing this concept, Waksman began a search for chemotherapeutic agents in soil microbes.

In 1944 his graduate students Albert Schatz and Elizabeth Bugie isolated *Actinomyces* (*streptomyces*) *griseus* from the throat of an infected chicken. From this fungus they extracted a new antimicrobial agent, effective against gram-negative bacteria. Waksman showed that it killed *Mycobacterium tuberculosis* too and named the compound "streptomycin." He called it an 'antibiotic.'

Waksman supplied ten grams of streptomycin to William Feldman and H. Corwin Hinshaw at the Mayo Clinic, who confirmed its anti-tuberculous properties—the news electrified the medical world. Soon, large clinical trials from Great Britain and elsewhere established streptomycin's efficacy against tuberculosis.

Waksman persuaded the Merck Company to manufacture streptomycin. Merck reaped huge profits from streptomycin sales and 80% of royalties went to Rutgers, 20% to Waksman. In 1949, following litigation settled out of court, Waksman shared the financial gains with 28 of his students, including Schatz.

Waksman was the first director of Rutgers' new Institute of Microbiology built from the streptomycin profits. Waksman remained active all his life, writing over 20 books, including *My Life with the Microbes* (Simon and Schuster, 1954), and *The Conquest of Tuberculosis* (University of California Press, 1965).

References

1. Dowling HF. Fighting Infection: Conquests of the Twentieth Century. Cambridge, Mass.: Harvard University Press, 1977.
2. Waksman SA. (with Starkey R) Humus, 1936; Microbial Antagonisms and Antibiotic Substances, 1945; The Literature on Streptomycin, 1948-1952; The Actinomycetes, 1950-1966; Actinomycetes and Their Antibiotics, 1953 (with Lechevalier H); Soil Microbiology, 1954.
3. Waksman SA. The Soil and the Microbe, 1931

4. Waksman SA. Enzymes, 1926 (with Davidson W). Principles of Soil Microbiology, 1927
5. Woodruff HB, ed. Scientific Contributions of Selman A. Waksman. New Brunswick, N.J.: Rutgers University Press, 1968.

1953

☆

Physics
FRITS (FREDERIK) ZERNIKE
for his demonstration of the phase contrast method, especially for his invention of the phase contrast microscope.

☆

Chemistry
HERMANN STAUDINGER
for his discoveries in the field of macromolecular chemistry.

☆

Physiology or Medicine
SIR HANS ADOLF KREBS
for his discovery of the citric acid cycle and
FRITZ ALBERT LIPMANN
for his discovery of co-enzyme, A and its importance for intermediary metabolism.

☆

Literature
SIR WINSTON LEONARD SPENCER CHURCHILL
for his mastery of historical and biographical description as well as for brilliant oratory in defending exalted human values.

☆

Peace
GEORGE CATLETT MARSHALL, General, President
American Red Cross, ex-Secretary of State and of Defense, Delegate to the U.N., Originator of the *Marshall Plan*.

1953

SIR HANS ADOLF KREBS (1900-81) AND FRITZ ALBERT LIPMANN (1899-1986)

Sir Hans Adolf Krebs   **Fritz Albert Lipmann**
© The Nobel Foundation

Sir Hans Adolf Krebs (left) and **Fritz Albert Lipmann** received the Nobel Prize in Physiology or Medicine "for his discovery of citric acid cycle" and Fritz Lipmann "for his discovery of coenzyme A."

Krebs was born in Hildesheim, Hanover. In 1925 he graduated in medicine from the University of Munich. Chance contacts brought him a job under the eminent biochemist Otto Warburg, from whom he learned tissue slice methods for metabolic research.

Carrying a heavy clinical workload in a Frieburg hospital, Krebs pursued research in biochemistry. In 1931 he discovered that ornithine increased urea output in liver tissue slices in presence of the ammonia and showed that ornithine acted catalytically. He proposed a cyclical process of ornithine, citrulline, and arginine in urea production.

Anti-Semitism forced Krebs out of Germany; arriving penniless at Cambridge in 1933, he soon resumed his research career and joined the University of Sheffield in 1935. His work there led him to discover that the final catabolic steps in glucose breakdown was also a cyclical process. He identified numerous related processes in a cyclical form and named it the "citric acid cycle," often called *The Krebs Cycle.*

Lippmann was born in Königsberg in East Prussia and studied medicine in Berlin. Feeling "uneasy about charging people for trying to make them healthy," he switched from medical practice to biochemistry research, and worked under Otto Meyerhof in Berlin. While studying pyruvate metabolism in bacteria, Lippmann observed that inorganic phosphate was a key ingredient for oxidation. In 1945 he identified (and named) "coenzyme A," as a cofactor during transfer of energy-rich bonds. CoA proved to be a derivative of the vitamin pantothenic acid.

Like Krebs, Lippmann was also a victim of anti-Semitism. He emigrated to the United States in 1939 and settled there. Colleagues and students deeply admired both scientists for their exemplary character and intellectual honesty. Asked about his work ethics, Krebs once quoted Noël Coward: "Work is fun, and there is no fun like work."

References

1. Holmes FL. Hans Krebs and the Discovery of the Ornithine Cycle. Federation Proceedings 39, pp. 216-225, February 1980.
2. Krebs SHA (with W.A. Johnson). The Role of Citric Acid in Intermediary Metabolism in Animal Tissues. Enzymologia, vol. 4, pp. 148-156, 1937.
3. Krebs SHA The Intermediary Stages in the Biological Oxidation of Carbohydrate. Advances in Enzymology, vol. 1, pp. 99-127, 1941.
4. Lipmann FA (with Levene PA). Serine Phosphoric Acid Obtained

on Hydrolysis of Vitellinic Acid. Journal of Biological Chemistry, vol. 98, pp. 109-117, 1932.

5. Lipmann FA Metabolic Generation and Utilization of Phosphate Bond Energy. Advances in Enzymology, vol. 1, pp. 99-162, 1941.

1954

☆

Physics
MAX BORN
for his fundamental research in quantum mechanics, especially
for his statistical interpretation of the wave function, and
WALTHER BOTHE
for the coincidence method and his discoveries made therewith.

☆

Chemistry
LINUS CARL PAULING
for his research into the nature of the chemical bond and its application
to the elucidation of the structure of complex substances.

☆

Physiology or Medicine
JOHN FRANKLIN ENDERS, THOMAS HUCKLE
WELLER and FREDERICK CHAPMAN ROBBINS
for their discovery of the ability of poliomyelitis viruses
to grow in cultures of various types of tissue.

☆

Literature
ERNEST MILLER HEMINGWAY
for his mastery of the art of narrative, most recently
demonstrated in The Old Man and the Sea *and for the*
influence that he has exerted on contemporary style.

☆

Peace
OFFICE OF THE UNITED NATIONS HIGH
COMMISSIONER FOR REFUGEES
An international relief organization founded
by the United Nations in 1951.

1954

John Franklin Enders (1897-1985), Thomas Huckle Weller (born 1915-), and Frederick Chapman Robbins (born 1916-)

John Franklin Enders   **Frederick Chapman Robbins**   **Thomas Huckle Weller**

© The Nobel Foundation

John Franklin Enders (left), **Thomas Huckle Weller** (bottom), **and Frederick Chapman Robbins** shared the Nobel Prize in Physiology or Medicine for work on poliomyelitis.

Enders was born in West Hartford, Connecticut. After studying literature at Harvard, he switched to biology and obtained a doctorate in this subject. In 1929, he joined Harvard's Bacteriology and Immunology department.

Weller was born in Ann Arbor, Michigan, and studied medicine at Harvard. His roommate there was Frederick Robbins, a youngster from Auburn, Alabama. Both obtained their MD degree in 1940, but interrupted residency training because of military service in the US Army during World War II. After the war, they joined the new infectious disease laboratory headed by Enders at Boston's Children's Hospital.

For years Enders had been working on culturing herpes simplex, mumps and measles viruses, with mixed results. Starting in 1947, he enlisted the assistance of Weller and Robbins and succeeded in maintaining viral growth for 30 days or longer in tissue cultures by repeated replenishment of the nutrient media. Postulating that specific viruses might grow better in their natural host cells he began using human embryonic skin and muscle tissues and successfully cultured chicken pox virus.

But culturing poliovirus in extra-neural tissues posed major hurdles. The Boston team serendipitously discovered that like varicella, the Lansing strain of the poliovirus could be cultured in human embryonic tissues using a 'roller tube method.' They also proved that the cultured poliovirus can be transmitted to monkeys via non-neural routes. The newly discovered antibiotics, penicillin and streptomycin, further aided in maintaining viral cultures free from bacterial contamination.

These pivotal discoveries helped decipher the epidemiology of poliomyelitis. The first extensive field trial of immunization with polio vaccine for half a million American school children was completed in the spring and summer of 1954. While poliomyelitis remains endemic in isolated areas, it is so well controlled worldwide, that generations of children have grown without the fear of this devastating disease.

References

1. Calder R. Man's Struggle Against Poliomyelitis. World Health 14, pp. 29-35, 1961.
2. Enders JF (with Hammon WD). Virus Disease of Cats, Principally Characterized by Aleucytosis, Enteric Lesions, and Presence of Intranuclear Inclusion Bodies. Journal of Experimental Medicine, vol. 69, pp. 327-352, 1939.

3. Enders with Zinsser H, Fothergill L: Immunity: Principles and Application in Public Health, 1939.

4. Paul JR. A History of Poliomyelitis. New Haven, Conn.: Yale University Press, 1971.

1955

☆

Physics
WILLIS EUGENE LAMB
for his discoveries concerning the fine structure of the hydrogen spectrum, and
POLYKARP KUSCH
for his precision determination of the magnetic moment of the electron.

☆

Chemistry
VINCENT DU VIGNEAUD
for his work on biochemically important sulphur compounds,
especially for the first synthesis of a polypeptide hormone.

☆

Physiology or Medicine
AXEL HUGO THEODOR THEORELL
for his discoveries concerning the nature and
mode of action of oxidation enzymes.

☆

Literature
HALLDÓR KILJAN LAXNESS
for his vivid epic power which has renewed the great narrative art of Iceland.

☆

Peace
No Award

1955

AXEL HUGO THEODOR THEORELL (1903-1982)

© The Nobel Foundation

Hugo Theorell was awarded the Nobel Prize in Physiology or Medicine "for discoveries relating to the nature and mode of action of oxidizing enzymes."

He was born in Linköping, Sweden, where his father was a surgeon and mother, an accomplished pianist and singer. She was the one who instilled in Hugo a lasting love of music.

In 1930, Theorell graduated in medicine from the Karolinska Institute in Stockholm and began research there. Working on muscle pigments, Theorell first characterized myoglobin and proved that it was a different chemical from hemoglobin. A Rockefeller Fellowship in 1933 helped him work under the famous biochemist Otto Warburg in Berlin. There, Theorell studied the nature of the oxidation enzyme, the "yellow pigment," and demonstrated that it contained a colorless protein fraction, and a colored coenzyme fraction related to riboflavin, or vitamin B2; this was later identified as flavin-mononucleotide. Theorell became the first scientist to isolate a coenzyme. Warburg called him, the "Master of Enzyme Research."

Returning to Sweden in 1935 Theorell assumed the directorship of the new Biochemical Department of the Nobel Medical Institute. Sweden's neutrality during World War II allowed Theorell to continue research uninterrupted. With colleagues, he purified cytochrome *c*—the 'red substance' known since 1886—and studied the velocities of enzyme reactions. The work of Theorell and associates thus became central to the advancement of molecular biochemistry of energy metabolism.

When he was seven years old, Theorell suffered a crippling attack of poliomyelitis. Although he walked unaided after a muscle transplantation in later years, his physical activity was much curtailed. But with his infectious enthusiasm toward life and brilliant intellect he rose far above his handicap. Besides becoming an outstanding scientist of this century, Theorell was an adept amateur musician.

References

1. Huxley AF. Energy Transformations in Biological Systems. Amsterdam: Elsevier, 1975.
2. Keilin D. The History of Cell Respiration and Cytochrome. Prepared for publication by Joan Keilin. Cambridge, England: Cambridge University Press, 1966.
3. Theorell AHT (with Åke Åkeson). Studies of Cytochrome c: I, Electrophoretic Purification of Cytochrome c and Its Amino Acid Composition, II, The Optical Properties of Pure Cytochrome c and Some of Its Derivatives, III, Titration Curves and IV, the Magnetic Properties of Ferric and Ferrous Cytochrome c. Journal of the American Chemical Society, vol. 63, pp. 1804-1827.
4. Theorell AHT Heme-Linked Groups and Mode of Action of Some Hemoproteins. Advances in Enzymology, vol. 7, pp. 265-303, 1947.

1956

☆

Physics
WILLIAM SHOCKLEY, JOHN BARDEEN
and **WALTER HOUSER BRATTAIN**
for their research on semiconductors and their discovery of the transistor effect.

☆

Chemistry
SIR CYRIL NORMAN HINSHELWOOD and
NIKOLAY NIKOLAEVICH SEMENOV
for their research into the mechanism of chemical reactions.

☆

Physiology or Medicine
ANDRÉ FRÉDÉRIC COURNAND, WERNER
FORSSMANN and **DICKINSON W. RICHARDS**
*For their discoveries concerning heart catheterization and
pathological changes in the circulatory system.*

☆

Literature
JUAN RAMÓN JIMÉNEZ
*for his lyrical poetry, which in Spanish language constitutes
an example of high spirit and artistic purity.*

☆

Peace
No Award

1956

WERNER FORSSMANN (1904-1979),
ANDRÉ FRÉDÉRIC COURNAND (1895-1988),
AND DICKINSON WOODRUFF
RICHARDS, JR. (1895-1973)

André Frédéric   Werner Forssmann   Dickinson Woodruff
Cournand   Richards Jr

© The Nobel Foundation

André Cournand (left), **Werner Forssmann** (right)**, and Dickinson Richards** shared the Nobel Prize in Physiology or Medicine for discoveries concerning heart catheterization and circulatory physiology.

Forssmann was born in Berlin. He graduated in medicine in 1929 and joined the Eberswalde Surgical Clinic. During his internship Forssmann wanted to administer drugs directly into the heart by performing cardiac catheterization; his superiors however, flatly denied permission for such an evidently risky procedure. Forssmann then practiced the procedure in cadavers, and with the help from an 'unsuspecting' nurse secretly tried cardiac catheterization on himself. He injected a local anesthetic, punctured a vein on his forearm and introduced a "well oiled" ureteral catheter. He pushed the catheter 65 cm toward his heart, walked downstairs to the x-ray room and under fluoroscopy, located the catheter tip. In his 1929 paper, "probing of the heart," he described how safe the procedure was. The report was sensational, but it also drew intense criticism from German doctors. They deplored him for performing a "dangerous stunt." Forssmann abandoned cardiology and trained in urology.

In the early 1930s, Dickinson Richards, an internist in New York's Bellevue Hospital, had been studying pulmonary functions in collaboration with André Cournand, a Parisian who had emigrated to the United States. To measure the mixed venous oxygen content these scientists had to obtain blood from the right atrium. They improvised Forssmann's catheterization technique and carried it out in animals. After four years of studies, they performed the first human cardiac catheterization in the USA and showed its safety and utility. The procedure was soon widely adopted, paving the way for major advances in cardiopulmonary physiology, clinical cardiology, and heart surgery.

During World War II, Forssmann had served as a medical officer in the SS Army—a record that was to haunt him all his life. In 1945 he was imprisoned for six months as an allied prisoner-of-war. Later, until the announcement of the 1956 Nobel Prizes, he had been practicing urology in relative anonymity in a small German town. Known for his delightful humor, it seems that Forssmann remarked that he had been away so long from academic milieu and clinical cardiology that now he was a "leading fossil."

References

1. Forssmann W. Experiments on Myself: Memoirs of a Surgeon in Germany, 1974.
2. Richards DW. Medical Priesthoods and Other Essays, 1970.
3. Richards DW (with Fishman AP). Circulation of the Blood: Men and Ideas, 1964.

1957

☆

Physics
CHEN NING YANG and **TSUNG-DAO LEE**

for their penetrating investigation of the so-called parity laws which has led to important discoveries regarding the elementary particles.

☆

Chemistry
LORD ALEXANDER R. TODD

for his work on nucleotides and nucleotide co-enzymes.

☆

Physiology or Medicine
DANIEL BOVET

for his discoveries relating to synthetic compounds that inhibit the action of certain body substances, and especially their action on the vascular system and the skeletal muscles.

☆

Literature
ALBERT CAMUS

for his important literary production, which with clear-sighted earnestness illuminates the problems of the human conscience in our times.

☆

Peace
LESTER BOWLES PEARSON

former Secretary of State for External Affairs of Canada, President 7th Session of the United Nations General Assembly.

1957

DANIEL BOVET (1907-1992)

© The Nobel Foundation

Daniel Bovet was awarded the Nobel Prize in Physiology or Medicine "for discoveries relating to synthetic compounds designed to block the body's own chemicals."

Born in Neuchâtel, Switzerland, Bovet studied at the University of Geneva and in 1929, received a doctorate in biochemistry. He worked at the Pasteur Institute in Paris, where he met and married a coworker, Filomena Nitti, the daughter of Francesco Nitti—the former premier of Italy who had been exiled by the fascists. The Bovet couple worked together in all their research. In 1947 they moved to Rome, where Daniel Bovet took up Italian citizenship and became director of the Italian Health Institute.

The central theme of Bovet's research was exploring the relationship between the molecular structure of endogenous chemicals and their biological actions. Using this knowledge Bovet would then try synthesizing drugs for specific clinical conditions. Bovet's early research led him to the discovery that the active ingredient in the antimicrobial Prontosil discovered by Gerhard Domagk (1939 Nobel Laureate) was indeed sulfanilamide. This finding was pivotal for large-scale manufacturing of new classes of inexpensive sulfa drugs.

In 1937, Bovet and his colleague Ms. A.M. Staub began developing agents for blocking endogenous amines such as histamine. Within a few years they introduced pyrilamine compounds—mepyramine, cyprohepatadine (Antergan) and pyrilamine maleate (Neo-antergan)—the first effective remedies for the common cold. These advances were to trigger research in neuropharmacology and synthesis of many neuropsychiatric drugs, such as chlorpromazine and trifluoperazine.

Prior to 1940, tubocurare, which had numerous side effects, was the only available muscle relaxant for surgical anesthesia. A visit to Brazil in the 1940s stimulated Bovet to explore the chemistry of nerve poisons used for hunting by local Indians. Bovet extracted curare and developed gallamine and succinylcholine—synthetic muscle relaxants, less toxic than tubocurare. The new agents were soon incorporated into the practice of surgical anesthesia.

A man of broad humanistic vision and deep social convictions, Bovet implored scientists to focus their concerns beyond the narrow confines of their laboratories and help solve problems arising from poverty, ignorance, and social injustice.

References

1. Bovet D. Genetic Aspects of Learning and Memory in Mice. Science, vol. 163, pp. 139-149, 1969.
2. Bovet D. Introduction to Antihistaminic Agents and Antergan Derivatives. Annals of the New York Academy of Sciences, vol. 50, pp. 1089-1126, 1950
3. Bovet D. Some Aspects of the Relationship Between Chemical Constitution and Curare-Like Activity. Annals of the New York Academy of Sciences, vol. 54, pp. 407-432, 1951.

4. Bovet D. Synthetic Inhibitors of Neuromuscular Transmission, Chemical Structures, and Structure Activity Relationships. Neuromuscular Blocking and Stimulating Agents, vol. 1. pp. 243-294, 1972.

1958

Physics
PAVEL ALEKSEYEVICH CHERENKOV and
IGOR YEVGENYEVICH TAMM
for the discovery and the interpretation of the Cherenkov effect.

☆

Chemistry
FREDERICK SANGER
for his work on the structure of proteins, especially that of insulin.

☆

Physiology or Medicine
GEORGE WELLS BEADLE and EDWARD LAWRIE TATUM
for their discovery that genes act by regulating definite chemical events and
JOSHUA LEDERBERG
*for his discoveries concerning genetic recombination and the
organization of the genetic material of bacteria.*

☆

Literature
BORIS LEONIDOVICH PASTERNAK
*for his important achievement both in contemporary lyrical poetry
and in the field of the great Russian epic tradition. (Accepted first,
later caused by the authorities of his country to decline the prize.)*

☆

Peace
GEORGES HENRI PIRE
Father of the Dominican Order, Leader of the relief organization
for refugees, l'Europe du Coeur au Service du Monde.

1958

GEORGE WELLS BEADLE (1903-89),
EDWARD LAWRIE TATUM (1909-75) AND
JOSHUA LEDERBERG (BORN 1925)

George Wells Beadle   **Edward Lawrie Tatum**   **Joshua Lederberg**
© The Nobel Foundation

George Wells Beadle (left), **Edward Lawrie Tatum** (right), and **Joshua Lederberg** shared the 1958 Nobel Prize in Physiology or Medicine for their contributions to the field of genetics.

George Beadle was born in Wahoo, Nebraska. A rural upbringing had kindled in him the love of nature and an interest in biology. He studied at Cornell University and moved to Stanford University in 1937, where he invited Edward Tatum to join the research team; Tatum was from Boulder, Colorado and a biochemistry graduate from University of Wisconsin. Tatum moved to Yale University in 1945; the next summer, Lederberg, then a medical student, did research under him. Lederberg was so impressed with the experience that he quit medical school and pursued a research career under Tatum.

Thomas Hunt Morgan had shown that chromosomes transmitted genetic traits, but the mechanism of transmission was not known. Beadle and Tatum studied the biochemistry of sex-linked eye-color mutations in the fruit fly *Drosophilae melanogastor*. Their work on the larvae led them to postulate that the ultimate eye-color would depend upon the chemicals released from the influence of genes. To prove the theory, they chose the well-characterized fast-growing fungus *Neurospora crassa,* a common bread mold.

They prepared over 1000 single-spore larval cultures and induced mutations in them using x-rays. By manipulating nutrients in the culture media, they identified mutants secondary to several specific nutritional deficiencies, including vitamins B_6 and B_1. These findings led them to propose one of the profound concepts in genetics—the theory of "single gene-single enzyme."

In the mid-1940s, Lederberg and Tatum made another breakthrough; working on *E. coli* and other microbes they showed that the biochemical and genetic make-up of bacteria were fundamentally like those of higher animals.

Beadle became chancellor, later president, of the University of Chicago. With his wife Muriel, he wrote *The Language of Life* (Doubleday 1966). It was Tatum who coined the phrase 'biological engineering,' which he used in his Nobel Lecture and prophesized that this field would greatly benefit mankind. Besides his pioneering contributions to genetics, Lederberg served as consultant for the NASA Viking Space Project.

References

1. Beadle GW. Development of Eye Colours in Drosophila: Pupal Transplants and the Influence of Body Fluid on Vermilion. Proceedings of the Royal Society of London, series B, vol. 122, pp. 98-105, 1937.
2. Beadle GW. Genetics and Modern Biology. Philadelphia: American Philosophical Society, 1963.
3. Beadle GW. The Genes of Men and Molds. In Facets of Genetics:

Readings from Scientific American, edited by Adrian Morris Srb. San Francisco: W.H. Freeman, 1970.

4. Beadle M. Where Has All the Ivy Gone? A Memoir of University Life, Garden City, N.Y. Doubleday, 1972.

5. Hayes W. The Genetics of Bacteria and Their Viruses. New York: John Wiley & Sons, 1964.

6. Lederberg J (with Tatum EL). Gene Recombination in Escherichia Coli. Nature, vol. 158, p. 558, 1946

7. Lederberg J, Tatum EL. Annual Review of Genetics 13, pp. 1-5, 1979.

8. Lederberg J. Genetic Exchange in Salmonella. Journal of Bacteriology, vol. 64, pp. 679-699, 1952

9. Lederberg J. Prospects for the Genetics of Somatic and Tumor Cells. Annals of the New York Academy of Science, vol. 63, pp. 662-665, 1956.

10. Lederberg J. Sex in Bacteria: General Studies, 1945-1952. Science, vol. 118, pp. 169-175, 1954

11. Tatum EL. Vitamin B Requirements of Drosophila melanogaster. Proceedings of the National Academy of Sciences, vol. 27, pp. 193-197, 1941.

12. Tatum EL. Genetic Control of Biochemical Reactions in Neurospora. Proceedings of the National Academy of Sciences, vol. 27, pp. 499-506, 1941.

1959

☆

Physics
EMILIO GIO SEGRÈ and **OWEN CHAMBERLAIN**
for their discovery of the antiproton.

☆

Chemistry
JAROSLAV HEYROVSKY
for his discovery and development of the polarographic methods of analysis.

☆

Physiology or Medicine
SEVERO OCHOA and **ARTHUR KORNBERG**
*for their discovery of the mechanisms in the biological synthesis
of ribonucleic acid and deoxyribonucleic acid.*

☆

Literature
SALVATORE QUASIMODO
*for his lyrical poetry, which with classical fire expresses
the tragic experience of life in our own times.*

☆

Peace
PHILIP J. NOEL-BAKER
Member of Parliament Great Britain; life-long ardent
worker for international peace and co-operation.

1959

SEVERO OCHOA (1905-1993) AND ARTHUR KORNBERG (BORN 1918)

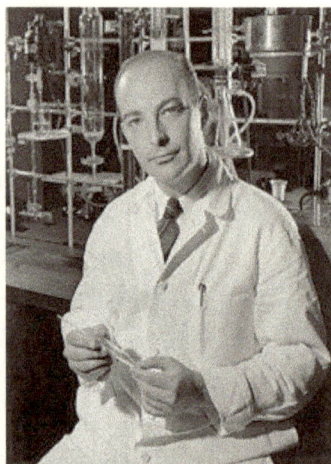

Severo Ochoa   **Arthur Kornberg**
© The Nobel Foundation

Severo Ochoa (right) and **Arthur Kornberg** shared the Nobel Prize in Physiology or Medicine "for their discovery of the mechanisms in the biological synthesis of RNA and DNA."

Son of a lawyer, Ochoa was born in Luarca, Spain and studied medicine at the University of Madrid. He was inspired by the works of the great Spanish neurologist Santiago Ramon y Cajal (the 1906 Nobel Laureate), who also had studied in the same school. In 1940 Ochoa moved to the United States and worked under Carl and Gerty Cori at Washington University in St. Louis. Later, Ochoa joined the faculty at New York University College of Medicine.

The son of Joseph and Lena (née Katz) Kornberg, Arthur was born in Brooklyn, New York. His parents had immigrated from Austria and his father operated sewing machines in sweatshops prior to owning a small hardware store. Arthur was a brilliant student with a reputation as the "smart kid on the block." His love of biology and biochemistry was sparkled after he took a premedical course at the City College of New York. Enrolling himself for medical studies at Rochester University, New York, Kornberg earned his medical degree from there in 1941.

After serving in the US Coast Guard, Kornberg worked at the nutrition division of the National Institutes of Health (NIH) from 1942-45 There he became the chief of the enzyme and metabolism section between 1947-53. Between 1941-53, he took leaves periodically, to do research with Severo Ochoa at the New York University College of Medicine, with Carl and Gerty Cori at Washington University School of Medicine in St. Louis, and with H. A. Baker at the University of California, Berkley. Since 1959, he stayed at Stanford University in California.

In 1955 Ochoa and Grunberg-Manago isolated a new enzyme from *Azobacter vinelandii* that was capable of synthesizing RNA in test tubes. They named the enzyme polynucleotide phosphorylase. Some years later it was shown that polynucleotides synthesized in-vitro were active as messengers in protein synthesis. Working independently, in 1953, Kornberg overcame technical limitations in purification, and isolated from *E. coli*, the DNA polymerase enzyme; using this enzyme one could synthesize DNA that was virtually identical to the natural product.

Thus, nearly 100 years after the discovery of nucleic acids, DNA and RNA could be artificially synthesized. About their work Kornberg remarked that they "had opened up a tiny crack and tried driving a wedge—the hammer was enzyme purification."

A prolific writer himself, Kornberg knew the difficulties of good writing, which he referred to as a 'variety of mental torture.' His autobiography *For the Love of Enzymes: The Odyssey of a Biochemist* was published in 1989 (Harvard University Press).

References

1. Kornberg A. American Men and Women of Science: Physical and Biological Sciences. 8 vols. 17th ed. New York: R.R. Bowker, 1989.
2. Kornberg A. Enzymatic Synthesis of DNA, 1962; Biosynthesis of DNA, 1964; DNA Synthesis, 1974; DNA Replication, 1980.
3. Ochoa S. Biosynthesis of Tricarboxylic Acids by Carbon Dioxide Fixation: I, Preparation and Properties of Oxalosuccinic Acid. Journal of Biological Chemistry, vol. 174, pp. 115-122, 1948; II, (with Tabori EW) Oxalosuccinic Carboxylase. Journal of Biological Chemistry, vol. 174, pp. 123-132, 1948; III, Enzymatic Mechanisms. Journal of Biological Chemistry, vol. 174, pp. 133-172, 1948.

1960

☆

Physics
DONALD A. GLASER
for the invention of the bubble chamber.

☆

Chemistry
WILLARD FRANK LIBBY
*for his method to use carbon-14 for age determination in
archaeology, geology, geophysics, and other branches of science.*

☆

Physiology or Medicine
SIR FRANK MACFARLANE BURNET and
SIR PETER BRIAN MEDAWAR
for discovery of acquired immunological tolerance.

☆

Literature
SAINT-JOHN PERSE (penname of ALEXIS LÉGER)
*for the soaring flight and the evocative imagery of his poetry which
in a visionary fashion reflects the conditions of our time.*

☆

Peace
ALBERT JOHN LUTULI
*President of the South African liberation movement,
the African National Congress.*

1960

SIR FRANK MACFARLANE BURNET (1899-1985), AND SIR PETER BRIAN MEDAWAR (1915-87)

Sir Frank Macfarlane Burnet   **Sir Peter Brian Medawar**

© The Nobel Foundation

Macfarlane Burnet (left), and **Peter Medawar** won the 1960 Nobel Prize in Physiology or Medicine "for their discoveries concerning acquired immunological tolerance."

Burnet was born in Traralgon, Victoria, in Australia, and studied medicine at the University of Melbourne. After a Ph.D. from London in 1927, he returned to Australia, but revisited England in 1932 and trained in virology at London's National Institute for Medical Research. Here he developed viral culture techniques that are being used even today, particularly for influenza virus. Virology remained Burnet's passion all his life.

Son of a naturalized British subject, Medawar, was born in Rio de Janeiro and attended the Magdalen College at Oxford. He taught in Oxford, Birmingham and London, and in 1962 became the director of the National Institute for Medical Research in London.

In 1945 Ray David Owens had discovered that cattle twins maintain a 'stable' mixture of each other's RBCs throughout their lives, which he thought was due to an intrauterine transfer of erythroid stem cells. Conceptualizing the significance of this discovery, Burnet proposed the "clonal selection theory," according to which, immunological tolerance was explained by the body's ability to recognize only those tissues present before a critical period during embryonic development as "self" and all others as "non-self."

Medawar's laboratory provided the proof for Burnet's hypothesis. Medawar injected spleen cells from one strain of mouse into another during the latter's embryonic period and demonstrated that as adults, the recipients showed tolerance for the donors' skin grafts. He proved that the rejection phenomenon was like 'cellular hypersensitivity' and argued that cell-mediated and antibody-mediated immunity belonged to different systems. Although immunology has greatly advanced since their time, Burnet and Medawar were pioneers who launched this subject on a modern path.

Both men were of exceptional intellect and deep social vision. Medawar's books on science, society and professional ethics remain classics—as for example his, 1961 book *The Strange Case of Spotted Mice and Other Classic Essays in Science* dealing with fraud in research, and his 1986 autobiography, *Memoir of a Thinking Radish*.

References

1. Bibel DJ, ed. Milestones in Immunology: A Historical Exploration. Madison, Wis.: Science Tech, 1988.
2. Burnet SM. Biological Aspects of Infectious Disease, 1940 (also as Natural History of Infectious Disease, 1953)
3. Burnet SM. Changing Patterns: An Atypical Autobiography, 1968.

4. Burnet SM. The Clonal Selection Theory of Acquired Immunity, 1959.

5. Burnet SM. The Production of Antibodies: A Review and a Theoretical Discussion, 1941.

6. Medawar PB. Memoir of a Thinking Radish. Oxford, Oxford University Press. 1986.

7. Medawar PB. Pluto's Republic. Oxford, Oxford University Press. 1982.

8. Medawar PB. The Use of Skin Grafting to Distinguish Between Monozygotic and Dizygotic Twins in Cattle. Heredity, vol. 5, pp. 379-397, 1951

9. Medawar PB. Aristotle to Zoos: A Philosophical Dictionary of Biology, Oxford, Oxford University Press 1983.

10. Medawar PB. The Strange Case of a Spotted Mice: and other Classic Essays in Science. Oxford, Oxford University Press. 1961.

11. Silverstein AM. A History of Immunology. San Diego: Academic Press, 1989.

12. Wintrobe MM. Blood, Pure and Eloquent: A Story of Discovery, People, and Ideas. New York: McGraw-Hill, 1980.

1961

☆

Physics
ROBERT HOFSTADTER

*for his pioneering studies of electron scattering in atomic nuclei
and for his thereby achieved discoveries concerning the structure of
the nucleons;* and **RUDOLF LUDWIG MÖSSBAUER**
*for his research concerning the resonance absorption of gamma radiation
and his discovery in this connection of the effect which bears his name.*

☆

Chemistry
MELVIN CALVIN

for his research on the carbon dioxide assimilation in plants.

☆

Physiology or Medicine
GEORG VON BÉKÉSY

for his discoveries of the physical mechanism of stimulation within the cochlea.

☆

Literature
IVO ANDRIC

*for the epic force with which he has traced themes and depicted
human destinies drawn from the history of his country.*

☆

Peace
DAG HJALMAR AGNE CARL HAMMARSKJÖLD

Secretary General of the United Nations
(awarded the Prize posthumously).

1961

GEORG VON BÉKÉSY (1899-1972)

© The Nobel Foundation

George Békésy was awarded the 1961 Nobel Prize in Physiology or Medicine for his discoveries related to physics of acoustics.

He was born in Budapest and studied in Berlin and Switzerland. After a Ph.D. in physics from the University of Budapest, he could not find work in his field of optics, forcing him to accept a job at the Hungarian Post and Telegraph.

The Hungarian telephone links were pivotal for transcontinental communications in the 1920s; but the lines often broke and locating the sites of disruption would take hours or days. Békésy reasoned that like musicians tuning their violins by plucking on the strings, locating the sites of disruption in the telephone lines should be possible by 'plucking' on the lines. By transmitting 'clicks' over the telephones, he refined a method of locating the sites of line disruptions.

Using the telephone model, Békésy studied auditory mechanics. He developed elegant techniques to measure the acoustics of the tympanic membrane, middle ear bones and the cochlea using laboratory animals and later, human ears. His findings proved that the frequency of the sound wave determined the sites of cochlear vibration, with higher frequencies resonating at the cochlear origin and lower frequencies, away from it. Békésy then showed that the nerve fibers attached to the cochlea transmit sounds as electrical signals of specific frequencies, aiding the brain to identify the pitch of the transmitted sound.

In 1947, Békésy emigrated to the United States and taught at Harvard and later at the University of Hawaii. Dedicated to his work, Békésy never married. He was an ardent art collector. On the eve of the Nobel ceremonies, it seems he joked that he needed another Nobel Prize, because there was a Rembrandt for sale in Stockholm, the price of which was about twice the Nobel Prize. In his will, he bequeathed all his invaluable collections to the Nobel Foundation in gratitude for the honor it had bestowed upon him.

References

1. Békésy G. Concerning the Pleasures of observing and the mechanisms of the inner ear. Nobel Lecture. Stockholm. 1961.
2. Békésy G. Experiments in Hearing. Translated and edited by E.G. Wever. New York, McGraw Hill. 1960
3. Békésy G. Sensory Inhibition, Princeton, Princeton University Press 1967.
4. Bernhard CG. Georg von Békésy and the Karolinska Institute *Hearing Research*. 22 (1986)13-17.
5. Tonndorf J. Georg von Békésy and his work. *Hearing Research*. 22 (1986) 3-10.
6. Wirgin J. The art collection of Georg von Békésy. *Hearing Research*. 22 (1986) 11-12.

1962

☆

Physics
LEV DAVIDOVICH LANDAU
for his pioneering theories for condensed matter, especially liquid helium.

☆

Chemistry
MAX FERDINAND PERUTZ and SIR
JOHN COWDERY KENDREW
for their studies of the structures of globular proteins.

☆

Physiology or Medicine
FRANCIS HARRY COMPTON CRICK, JAMES DEWEY
WATSON and MAURICE HUGH FREDERICK WILKINS
for their discoveries concerning the molecular structure of nuclear acids and its significance for information transfer in living material.

☆

Literature
JOHN STEINBECK
for his realistic and imaginative writings, combining as they do sympathetic humor and keen social perception.

☆

Peace
LINUS CARL PAULING
California Institute of Technology, Pasadena, CA.
Campaigner especially for an end to nuclear weapons tests.

1962

| Francis Harry Compton Crick | James Dewey Watson | Maurice Hugh Frederick Wilkins |

© The Nobel Foundation

Francis Crick (left), **James Watson** (right), and **Maurice Wilkins** were awarded the 1962 Nobel Prize in Physiology or Medicine for their discovery of the structure of DNA.

Crick was born in Northampton, England. After training in physics, he began working at the Cavendish Laboratory at Cambridge. In 1951, James Watson, a Chicagoan trained in zoology from the University of Indiana, joined Crick at Cambridge as a research fellow to work on nucleic acids.

Although nucleic acids had been known for fifty years, their structure and functions were unknown in the 1940s. But several scientists were on their trail, including the physicist and x-ray crystallographer Rosalind

Franklin (1920-58) and her colleague at King's College in London, Maurice Wilkins — a biophysicist from New Zealand. By 1951 Franklin had photographed the DNA molecule and identified the location of its phosphate sugars. Her notebook entries suggests that she had thought the DNA to be a 'spiral.'

During a conference in Naples in May 1951, Watson attended a talk by Wilkins, in which the latter "flicked on the screen near the end of his talk" a photograph of DNA Franklin had made. Back in Cambridge, after a series of brilliant modeling studies, Watson and Crick proposed a double helix structure for DNA in their famous single-page paper (*Nature* 1953; 171:739-40). Implying that DNA carried the genetic code, they concluded the paper in a prophetic understatement: "It has not escaped our notice that the specific pairing we have postulated [for DNA] immediately suggests a possible copying mechanism for the genetic material."

Watson moved to the US in 1955 and later led the Human Genome Project. Between 1976 and 1995 Crick researched brain functions at California's Salk Institute. Wilkins became professor emeritus in London University. In 1958, Franklin died of ovarian cancer.

The elucidation of the structure of DNA was one the greatest scientific achievements of all time and a landmark in biology. Many historians believe that had she lived, Franklin should have shared the Nobel Prize for her part in this discovery.

References

1. Crick F (with Watson JD). Molecular Structure of Nucleic Acids: A Structure for Deoxyribose Nucleic Acid. Nature, vol. 171, pp. 737-738, 1953.
2. Crick F. Lessons from Biology. Natural History 97; pp. 32-39, November 1988.
3. Judson HF. The Eighth Day of Creation: The Makers of the Revolution in Biology. New York: Simon & Schuster, 1979.

4. Watson JD (with Tooze J). Molecular Biology of the Gene, 1965; The Double Helix, 1968; The DNA Story, 1981.

5. Wilkins MHF (with Stokes AR, Wilson HR). Molecular Structure of Deoxypentose Nucleic Acids. Nature, vol. 171, 737, 1953.

1963

☆

Physics
EUGENE P. WIGNER
*for his contributions to the theory of the atomic nucleus and
the elementary particles, particularly through the discovery and
application of fundamental symmetry principles;* and
MARIA GOEPPERT-MAYER and **J. HANS D. JENSEN**
for their discoveries concerning nuclear shell structure.

☆

Chemistry
KARL ZIEGLER, and **GIULIO NATTA**
*for their discoveries in the field of the chemistry
and technology of high polymers.*

☆

Physiology or Medicine
**SIR JOHN CAREW ECCLES, SIR ALAN LLOYD
HODGKIN** and **SIR ANDREW FIELDING HUXLEY**
*for their discoveries concerning the ionic mechanisms
involved in excitation and inhibition in the peripheral
and central portions of the nerve cell membrane.*

☆

Literature
GIORGOS SEFERIS (penname of **GIORGOS SEFERIADIS**)
*for his eminent lyrical writing, inspired by a deep
feeling for the Hellenic world of culture.*

☆

Peace
INTERNATIONAL COMMITTEE OF THE RED CROSS
Geneva, Founded 1863, and
LEAGUE OF RED CROSS SOCIETIES
Geneva

1963

SIR JOHN CAREW ECCLES (1903-1997),
SIR ALAN LLOYD HODGKIN (1914-1998),
AND SIR ANDREW FIELDING
HUXLEY (1917-2012)

Sir John Carew Eccles **Sir Alan Lloyd** **Sir Andrew**
                          Hodgkin   **Fielding Huxley**

© The Nobel Foundation

John Eccles (left), **Alan Hodgkin** (right), and **Andrew Huxley** shared the Nobel Prize in Physiology or Medicine for work on nerve transmission.

Born in Banbury, Oxfordshire, England, Alan Hodgkin studied science at Trinity College, Cambridge. His mentor Edgar D. Adrian advised Hodgkin to study nerve conduction in the large, unmyelinated, crab neurons because they were easy to prepare.

Huxley was born in London and educated at Trinity. The famous scientist Thomas Huxley was his grandfather and the celebrated writer Aldous Huxley, his half-brother. In 1939, Huxley began research collaboration with Hodgkin at the Marine Biological Laboratories in Plymouth, England. The onset of the Second World War interrupted their research, which they resumed soon after the war.

Using the "voltage-clamp" method, in which two microelectrodes are inserted into the nerve fibers to create a break in the impulse flow, they measured axon permeability, ion transport, and biophysical properties of nerve conduction. Applying this technique to the giant axons from squids (*Laligo*) and cuttlefish (*Sepia*), they discovered that energy-dependent Na^+ and K^+ pumps were present on the nerve fiber membranes.

Eccles was born in Melbourne and studied medicine there. In 1925 he went to Oxford University on a Rhodes Scholarship and worked under Sir Charles Sherrington. From 1937 through the 1960s, he led several research teams in Australia, New Zealand, and the United States, studying physiology of the central and peripheral nervous system.

Using electron microscopic and micro-puncture methods, Eccles and colleagues investigated the electrical and ionic changes at the synapses during nerve impulse transmission. Among other things, they discovered that synaptic junctions contained vesicles that released both excitatory as well as inhibitory neurotransmitters, and impulse transmission involved changes in electrical gradients and ionic composition in and around the axons. Thus, these findings complemented the discoveries of Hodgkin and Huxley.

Eccles was knighted in 1958, Hodgkin in 1972, and Huxley in 1974. In his later years, Eccles explored the relationship between the mind and the body. In his classic book, *The Self and Its Brain* (Springer-Verlag, 1977), written in collaboration with the philosopher Karl Popper, Eccles discussed the neuronal mechanisms of self-awareness and their relation to the physical and the cultural worlds made by man.

References

1. Cronin J. Mathematical Aspects of Hodgkin-Huxley Neural Theory. Cambridge, England: Cambridge University Press, 1987.
2. Eccles SJC. Excitatory and Inhibitory Synaptic Action. Annals of the New York Academy of Sciences, vol. 81, pp. 247-264, 1959

3. Eccles SJC. Facing Reality: Philosophical Adventures of a Brain Scientist, 1970; (with Popper K) The Self and Its Brain, 1977

4. Eccles SJC. Ionic Mechanism of Postsynaptic Inhibition. Science, vol. 145, pp. 1140-1147, 1964

5. Eccles SJC. The Neurophysiological Basis of Mind: The Principles of Neurophysiology, 1953.

6. Eccles SJC. (with Schmidt R, Willis WD) Pharmacological Studies on Presynaptic Inhibition. Journal of Physiology, vol. 168, pp. 500-530, 1963.

7. Eccles SJC. The Synapse. Scientific American, vol. 212, pp. 56-66.

8. Hodgkin AL. Beginning: Some Reminiscences of My Early Life (1914-1947). Annals Reviews in Physiology, vol. 45, pp. 1-16, 1983.

9. Hodgkin AL. The Conduction of the Nervous Impulse. Liverpool, England: Liverpool University Press, 1964.

10. Hodgkin AL. The Ionic Basis of Nervous Conduction. Science, vol. 145, pp. 1148-1154, September 11, 1964.

11. Huxley AF (with Hodgkin AL). Action Potentials Recorded from Inside a Nerve Fibre. Nature, vol. 144, p. 710 1939.

1964

☆

Literature
JEAN-PAUL SARTRE
*for his work which, rich in ideas and filled with the spirit of freedom and
the quest for truth, has exerted a far-reaching influence on our age.*
(Declined the prize.)

☆

Chemistry
DOROTHY CROWFOOT HODGKIN
*for her determinations by X-ray techniques of the
structures of important biochemical substances.*

☆

Physics
CHARLES H. TOWNES,
NICOLAY GENNADIYEVICH BASOV and
ALEKSANDR MIKHAILOVICH PROKHOROV
*for fundamental work in the field of quantum electronics, which has led to the
construction of oscillators and amplifiers based on the maser-laser principle.*

☆

Physiology or Medicine
KONRAD BLOCH and FEODOR LYNEN
*for their discoveries concerning the mechanism and regulation
of the cholesterol and fatty acid metabolism.*

☆

Peace
MARTIN LUTHER KING JR.
Leader of the Southern Christian Leadership
Conference, campaigner for civil rights.

1964

Konrad Emil Bloch   **Feodor Felix Konrad Lynen**
© The Nobel Foundation

Konrad Bloch (left), and **Feodor Lynen** shared the 1964 Nobel Prize in Physiology or Medicine for their independent work on the metabolism of cholesterol and fatty acids.

Bloch was born in Neisse, Germany (now in Poland), and studied chemical engineering and biochemistry initially in Munich and later in Switzerland. Because he was a Jew, he was denied reëntry into Germany; he emigrated to the United States, arriving in New York in 1936, with "great hopes and no money." Within 2 years, however, he earned a Ph.D. from Columbia University and spent the rest of his career in research.

Cholesterol had been isolated from gallstones in the 1820s and its structure was known since the 1930s. But neither its synthesis nor its metabolic fate was known. Bloch collaborated with David Rittenberg studying cholesterol metabolism. Injecting radioactive isotopes into rats, the team identified carbon and hydrogen isotopes being incorporated into cholesterol molecules. They characterized 36 biochemical steps involved in cholesterol synthesis and showed that acetic acid was its major precursor.

Son of a professor in mechanical engineering, Lynen was born in Munich and studied there. He survived the political turmoil in Munich in the mid-1930s and earned a Ph.D. in chemistry in 1937. Upon the recommendation of Otto Warburg, Lynen became the director of the Max Planck Institute for Cellular Chemistry in 1954.

Like Bloch, Lynen also was interested in lipid metabolism. In 1951, he discovered that acetyl coenzyme A initiated a series of reactions leading to the formation of cholesterol. Lynen too, independently elucidated all the 36 biochemical steps in cholesterol synthesis. In 1959, he discovered molecular functions of biotin by showing that this critical vitamin helped carbon dioxide incorporation during fat syntheses.

Today's advances in dietetic and medical approaches for cholesterol control and the treatment of cardiovascular diseases owe a great deal to the pioneering discoveries of Bloch and Lynen. Although each of them had worked independently, once they learned that their topics of research were similar, they immediately began corresponding and sharing data and results—a rare phenomenon among competing scientists.

References

1. Bloch KE (with Berg A, Rittenberg D). Biological Conversion of Cholesterol to Cholic Acid. Journal of Biological Chemistry, vol. 149, pp. 511-517, 1943
2. Bloch KE (with Langdon RG) Biosynthesis of Squalene. Journal of Biological Chemistry, vol. 200, pp. 129-134, 1953
3. Bloch KE. (with Rittenberg D) The Utilization of acetic acid for

Fatty Acid Synthesis. Journal of Biological Chemistry, vol. 154, pp. 311-312, 1944

4. Bloch KE Lipid Metabolism, 1960

5. Bloch KE. Enzymatic Synthesis of Monosaturated Fatty Acids. Accounts of Chemical Research, vol. 2, pp. 193-202, 1962

6. Bloch KE. Summing Up. Annual Review of Biochemistry, vol. 56, pp. 1-19, 1987

7. Lymen F (with Lane MD). The Biochemical Function of Biotin: VI, Chemical Structure of the Carboxylated Active Site of Propionyl Carboxylase. Proceedings of the National Academy of Sciences (USA), vol. 49, pp. 379-385, 1963

8. Lymen F. Biosynthesis of Fatty Acids. Proceedings of the Symposium on Drugs Affecting Lipid Metabolism, 1961

1965

Physics
SIN-ITIRO TOMONAGA, JULIAN SCHWINGER and RICHARD P. FEYNMAN

for their fundamental work in quantum electrodynamics, with deep-ploughing consequences for the physics of elementary particles.

Chemistry
ROBERT BURNS WOODWARD

for his outstanding achievements in the art of organic synthesis.

Physiology or Medicine
FRANÇOIS JACOB, ANDRÉ LWOFF and JACOUES MONOD

for their discoveries concerning genetic control of enzyme and virus synthesis.

Literature
MICHAIL ALEKSANDROVICH SHOLOKHOV

for the artistic power and integrity with which, in his epic of the Don, he has given expression to a historic phase in the life of the Russian people.

Peace
UNITED NATIONS CHILDREN'S FUND (UNICEF) New

York, founded by U.N. in 1946; an international aid organization.

1965

Jacques Lucien Monod (1910-1976)
André Michel Lwoff (1902-1994),
and François Jacob (1920-2013)

Jacques Lucien Monod   André Michel Lwoff   François Jacob

© The Nobel Foundation

Jacques Monod (left), **André Lwoff** (right), and **François Jacob** shared the 1965 Nobel Prize in Physiology or Medicine "for their discoveries concerning genetic control of enzyme and virus synthesis."

André Lwoff's was born in Ainay-le-Château, France. After obtaining his MD in 1927, he joined the faculty of the Pasteur Institute. In 1938, he became the head of that institute's 'Microbic Physiology' department.

François Jacob was born in Nancy, France, and studied medicine at the University of Paris. In 1950, he also joined the Pasteur Institute in Lwoff's department where Jacques Monod, another Parisian and biologist had been working as the Laboratory Director.

After a decade of work on the morphogenesis and the nutritional needs of ciliated parasites, in 1930, Lwoff began studying the nature of the symbiotic relationship between bacterial viruses (bacteriophage) and the lysogenic (host) bacteria. He and his colleagues showed that the trait of "peaceful coexistence" between the virus and the host (lysogeny) can be transmitted on to the next generation through chromosomal alterations. They also discovered that mutation induced by ultraviolet light transformed the noninfectious forms of bacteriophage into infectious forms.

Pursuing the concept that each enzyme had its own regulating gene, Monod, Jacob and colleagues studied biochemical genetics; they postulated the existence of 'operon', or a segment of the chromosome containing the regulatory gene. In their essay, "Genetic Regulatory Mechanisms in the Synthesis of Proteins," (*Journal of Molecular Biology* 1961; 3:318-356), they explained how DNA transcribed its 'message' to an RNA molecule (they called it the "messenger RNA"), directing protein synthesis.

During the Nazi occupation of France, Jacob fled to England, joined the Free French Army and fought in North Africa and Normandy. He was wounded in action and was later awarded *Croix de Guerre,* and the *Companion of the Liberation*, two of France's highest military honors. For their roles during the Resistance, Lwoff won the *Medal of Resistance* and Monod the *Croix de Guerre.* A friend of Albert Camus, Monod was also philosopher of science; in *Chance and Necessity* (Knopf, 1971), his international bestseller, he discussed the role of biology in the real world, and how random mutation supported Darwinian concepts of evolution.

References

1. Fuller W, ed. The Biological Revolution: Social Good or Social Evil? Garden City, N.Y.: Anchor/Doubleday, 1972.
2. Jacob F. Genetics Aspects of Lysogeny. The Chemical Basis of Heredity, edited by McElroy WD, Glass B, 1957.
3. Jacob F. Transduction of Lysogeny in E. coli. Virology, vol. 1, pp. 207-220, 1955 Jacob F.
4. Lewis J, ed. Beyond Chance and Necessity: A Critical Inquiry

into Professor Jacques Monod's. Chance and Necessity. Atlantic Highlands, N.J.: Humanities Press, 1974.

5. Lwoff A. Identification of Macromolecules in Escherichia coli. Biological Order, 1962.

6. Lwoff A (with Hunter S). Introduction to Biochemistry of Protozoa. Biochemistry and Physiology of Protozoa, vol. 1, p. 1-7, 1964.

7. Monod JL. Selected Papers in Molecular Biology, edited by André Lwoff and Agnes Ullmann, 1978.

1966

Physics
ALFRED KASTLER
*for the discovery and development of optical methods
for studying hertzian resonance in atoms.*

Chemistry
ROBERT S. MULLIKEN
*for his fundamental work concerning chemical bonds and the
electronic structure of molecules by the molecular orbital method.*

Physiology or Medicine
PEYTON ROUS
for his discovery of tumor inducing viruses, and
CHARLES BRENTON HUGGINS
for his discoveries concerning hormonal treatment of prostatic cancer.

Literature
SHMUEL YOSEF AGNON
*for his profoundly characteristic narrative art with
motifs from the life of the Jewish people.*
NELLY SACHS
*for her outstanding lyrical and dramatic writing, which
interprets Israel's destiny with touching strength.*

Peace
No Award

1966

FRANCIS PEYTON ROUS (1879-1970) AND CHARLES BRENTON HUGGINS (BORN 1901-1997).

Francis Peyton Rous   **Charles Brenton Huggins**
© The Nobel Foundation

Peyton Rous (left), and **Charles Huggins** received the 1966 Nobel Prize for their independent discoveries on cancers.

Huggins was born in Halifax, Nova Scotia, Canada and graduated in medicine from Harvard University in 1924. For 40 years, he worked as surgeon and researcher at the University of Chicago Medical School. In 1972, he became the Chancellor of Acadia University in his native Canada.

As a urologist, Huggins began studying diseases of the prostate gland. His choice of dogs for studies turned out be fortunate because dogs were the only animals that developed prostate cancers spontaneously.

Huggins found high concentrations of serum testosterone in dogs with prostate cancers and showed that orchiectomy led to tumor regression, proving that some types of cancers needed hormones to fuel their growth. Huggins then pioneered hormone therapy and surgery on testes, ovaries and the adrenal glands in patients with advanced cancers.

Renewed interest in cancer in the 1950s drew attention to the forgotten discoveries of Francis Peyton Rous. A medical graduate from the Johns Hopkins University in 1905, Rous joined the Rockefeller Institute in New York as a pathologist in 1909. The same year, a local farmer brought a Plymouth Rock chicken with a tumor for testing. Having determined the tumor to be a sarcoma, Rous discovered that it could be 'transferred' into healthy chicken by grafting tumor cells. Surprisingly, injecting cell-free filtrates from the tumor also led to sarcomas in healthy chickens; by 1914, Rous' laboratory had discovered three distinct types of avian sarcomas. Although he did not isolate specific agents, Rous postulated that the tumors were due to "filtrable agents," which were eventually identified as viruses—the Rous sarcoma virus.

For over 40 years, the scientific community remained skeptical of Rous' viral hypothesis of sarcoma. Meanwhile, collaborating with Joseph Turner, Rous developed an anticoagulant mixture of citrated physiological saline and glucose as blood preservative. Using this Rous-Turner solution, Oswald Robertson, Rous' colleague, initiated the world's first blood bank at the Belgian front during World War I in 1917. In the 1930s and 1940s, Rous made other major discoveries in cancers, warts and tumor-causing viruses. Octogenarian Rous was pleased that the Nobel Foundation finally recognized his cancer work, albeit 55 years late. Rous continued research through his 90th birthday.

References

1. Bryan WR. Peyton Rous. Science 154, pp. 364-365, 1966.
2. Cant G. Male Trouble. The New York Times Magazine 14, February 16, 1975.

3. Huggins CB. Studies on Prostatic Cancer I: Effect of Castration, Estrogen, and Androgen Injection on Serum Phosphatases in Metastatic Carcinoma of the Prostate. Cancer Research, vol. I, pp. 293-297, 1941.

4. Rous P On the Causation by Filterable Agents of Three Distinct Chicken Tumors. Journal of Experimental Medicine, vol. 18, pp. 52-59, 1914.

5. Rous P (with Murphy JB). Transmission of a Malignant New Growth by Means of a Cell-Free Filtrate. Journal of the American Medical Association, vol. 56, p. 198, 1911

6. Rous P The Challenge to Man of the Neoplastic Cell. Science, vol. 157, pp. 24-28, 1967.

7. Rous P. Viruses and Tumour Causation: An Appraisal of Present Knowledge. Nature, vol. 207, pp. 457-463, 1965.

1967

☆

Physics
HANS ALBRECHT BETHE

for his contributions to the theory of nuclear reactions, especially his discoveries concerning the energy production in stars.

☆

Chemistry
MANFRED EIGEN, RONALD GEORGE WREYFORD NORRISH and LORD GEORGE PORTER

for their studies of extremely fast chemical reactions, effected by disturbing the equilibrium by means of very short pulses of energy.

☆

Physiology or Medicine
RAGNAR GRANIT, HALDAN KEFFER HARTLINE and GEORGE WALD

for their discoveries concerning the primary physiological and chemical visual processes in the eye.

☆

Literature
MIGUEL ANGEL ASTURIAS

for his vivid literary achievement, deep-rooted in the national traits and traditions of Indian peoples of Latin America.

☆

Peace
No Award

1967

GEORGE WALD (1906-1997); RAGNAR A GRANIT (1900-1991) AND HALDAN KEFFER HARTLINE (1903-1983).

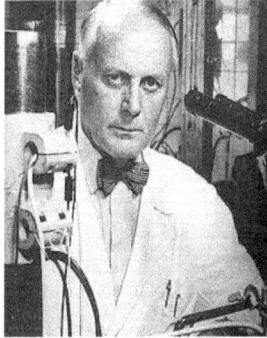

George Wald **Ragnar Arthur Granit** **Haldan Keffer Hartline**

© The Nobel Foundation

George Wald (left), **Ragnar Granit** (right), and **Haldan Hartline** shared the 1967 Nobel Prize in Physiology or Medicine for their discoveries related to the physiology of vision.

Granit was born in Helsinki, Finland, and obtained his medical degree there. After training in neuroscience under Sir Charles Sherrington in Oxford, Granit studied medical physics at the University of Pennsylvania. In 1940, he joined the Karolinska Institute in Stockholm, where he became the director of the Neurophysiology Institute.

Like Sherrington and Adrian, Granit used innovative microelectrodes to measure nerve impulses. He showed that light on the retina elicited both excitatory and inhibitory signals as measured in the optic nerves; the sum of these effects helped produce images perceived by the brain. Granit also worked on developing the electroretinogram (ERG), with which he and his colleagues studied the basic principles of color vision. ERG remains a major investigative tool in clinical and research ophthalmology.

Wald was born in New York City. After undergraduate studies, he joined the faculty at Harvard University's biology department. Studying the chemistry of the retina he found that rhodopsin was made up of opsin, a protein, and retinal, an aldehyde of vitamin A. He then showed that light caused a breakdown of the rhodopsin molecule and darkness reversed the process. These discoveries led Wald to postulate that a deficiency of vitamin A caused night blindness. In the 1950s, Wald and associates isolated other retinal pigments related to vitamin A and showed their role in color blindness.

Born in Bloomsburg, Pennsylvania and educated at Johns Hopkins University, Hartline began studying impulse transmission in the well-differentiated optic nerve fibers of horseshoe crabs. He discovered that signal rates were proportional to the intensity of light falling on the photoreceptors. In the frog visual system, he discovered that nerve fibers were activated according to the type of light, brightness, or object movement. Hartline also elucidated the mechanisms of retina signal processing, night-vision, and the recognition of patterns and motion by the retina.

Although many concepts formulated by Granit, Wald, and Hartline have changed, the research methods they developed and basic concepts they advanced were crucial for the growth of cognitive neuroscience and neuro-ophthalmology.

References

1. Barlow RB. What the Brain Tells the Eye. Scientific American 262, pp. 90-95, April 1990.
2. Granit R The Visual Pathway in The Eye, Ed. H. Davson, 1962.
3. Granit R. Sensory Mechanisms of the Retina, 1947.
4. Granit R. Muscular Afferents and Motor Control, 1966.
5. Granit R. The Development of Retinal Neurophysiology. Science, vol. 160, pp. 1192-1196, 1968.

6. Hartline HK. The Discharge of Nerve Impulses from the Single Visual Sense Cell. Cold Spring Harbor Symposia on Quantitative Biology, vol. 3, pp. 245-249, 1935.

7. Hartline HK. The Neural Mechanisms of Vision. The Harvey Lectures: 1941-1942, series 37, pp. 39-68, 1942.

8. Hubel DH. Eye, Brain and Vision. New York: Scientific American Library, 1988.

9. Stryer L. The Molecules of Visual Excitation. Scientific American 257, pp. 42-50, July 1987.

10. Wald G. Molecular Basis of Visual Excitation. Science, vol. 161, pp. 230-239, 1968.

11. Wald G. Defective Color Vision and its Inheritance. Proceedings of the National Academy of Sciences, vol. 55, pp. 1347-1363, 1966

12. Wald G. Eye and Camera. Scientific American, vol. 183, pp. 32-41, August 1950

13. Wald G. The Origin of Life. Scientific American, vol. 191, pp. 44-53, August 1954.

1968

☆

Physics
LARS ONSAGER
for the discovery of the reciprocal relations bearing his name, which are fundamental for the thermodynamics of irreversible processes.

☆

Chemistry
LUIS W. ALVAREZ
for his decisive contributions to elementary particle physics, in particular the discovery of a large number of resonance states, made possible through his development of the technique of using hydrogen bubble chamber and data analysis.

☆

Physiology or Medicine
ROBERT W. HOLLEY, HAR GOBIND KHORANA and MARSHALL W. NIRENBERG
for their interpretation of the genetic code and its function in protein synthesis.

☆

Literature
YASUNARI KAWABATA
for his narrative mastery, which with great sensibility expresses the essence of the Japanese mind.

☆

Peace
RENÉ CASSIN
President of the European Court for Human Rights.

1968

HAR GOBIND KHORANA (1922-2011), ROBERT HOLLEY (1922-1993) AND MARSHALL WARREN NIRENBERG (1927-2010)

Har Gobind Khorana   Robert William
Holley

Marshall Warren
Nirenberg

© The Nobel Foundation

Har Gobind Khorana (left), **Robert Holley** (right), **and Marshall Nirenberg** shared the 1968 Nobel Prize in Physiology or Medicine for their "interpretation of the genetic code and its function in protein synthesis."

Robert Holley was born in Urbana, Illinois and studied chemistry at the University of Illinois and at Cornell University. In 1957, he and his colleagues discovered a new type of RNA involved in protein synthesis, now known as the transfer RNA (tRNA). By developing catalytic enzyme techniques Holley then discovered the entire nucleotide sequence of alanine transfer RNA from yeast—the first such sequencing (*Science*, 1965; 147: 1462-65).

Khorana was born in Raipur, India (now in Pakistan) and studied chemistry in Lahore. He went to England on a scholarship from the

Indian Government and earned his Ph.D., from the University of Liverpool in 1948. In 1960 he joined the Institute for Enzyme Research at University of Wisconsin-Madison. Meanwhile, Nirenberg, a New Yorker with a doctorate in Zoology from the University of Michigan, Ann Arbor, had joined the National Institutes of Health to work on the biochemistry of genetics.

Nirenberg used cell-free filtrates, enzymes, and synthetic RNA polymers to study protein synthesis. He found that poly-uradilic acid (poly-U), an artificial mRNA containing exclusively the base uracil, stimulated the production of the polypeptide poly-phenylalanine. Thus was the first "word" of the genetic code deciphered: it was UUU coding for phenylalanine. With radiolabeled amino acids, artificial trinucleotides, and cellulose nitrate filters, Nirenberg further determined the exact codon alphabet of each of the amino acids.

Khorana extended Nirenberg's observations. By refining analytical techniques, Khorana and coworkers deciphered the genetic code for all the 64 triplet combinations of the four bases, including specific sequences for all the amino acids.

It is most remarkable that despite their diverse backgrounds and training, the contributions of these scientists melded so harmoniously, helping to understand the language of the gene. Of their work Maxine Singer said (*Science* 162, 1968,433-26) that "...these three men together constitute a triplet of great sense."

References

1. Celis JE, Smith JD, eds. Nonsense Mutations and tRNA Suppressors. New York: Academic Press, 1979.
2. Holley RW (with Baldwin JH, Greenfield S, Armour R). A Growth Regulatory Factor That Can Both Inhibit and Stimulate Growth. CIBA Foundation Symposium 116, pp. 241-252, 1985.
3. Holley RW. An Alanine-Dependent, Ribonuclease-Inhibited Conversion of AMP to ATP, and Its Possible Relationship to Protein

Synthesis. Journal of the American Chemical Society, vol. 79, pp. 658-662, 1957.

4. Khorana HG (with Matthaei H) Characteristics and Stabilization of DNAase-Sensitive Protein Synthesis in E. coli Extracts, Proceedings of the National Academy of Sciences, U.S.A., vol. 47, pp. 1580-1588, 1961.

5. Khorana HG. Studies on Bacteriorhodopsin and Rhodopsin. Proceedings of the Retina Research Foundation Symposium, vol. 1, pp. 62-89, 1988.

6. Khorana HG. Studies on Phosphorylation: Part XI, The Reaction Between Carbodiimides and Acids Esters of Phosphoric Acid. A New Method for the Preparation of Pyrophosphates. Journal of the Chemical Society, vol. 465, pp. 2257-2260, 1953

7. Khorana HG. The Chemistry of Carbodiimides. Chemical Reviews, vol. 53, p. 145, 1953; Khorana HG. Some Recent Developments in the Chemistry of Phosphate Esters of Biological Interest, 1961.

8. Khorana HG. Total Synthesis of a Gene. Science, vol. 203, p. 614, 1978

9. Nirenberg MW (with Matthaei H). The Dependence of Cell-Free Protein Synthesis in E. coli upon RNA Prepared from Ribosomes. Biochemical and Biophysical Research Communications, vol. 4, pp. 404-408, 1961

10. Olby R. The Path to the Double Helix. Seattle: University of Washington Press, 1974.

11. Portugal FH, Cohen JS. A Century of DNA: A History of the Discovery of the Structure and Function of the Genetic Substance. Cambridge, Mass.: MIT Press, 1977.

12. Schön A, Drupp G, Goug S, Kannangara CG, Söll D. A New Role for Transfer RNA: New Perspectives, edited by Masayori Inouye and Bernard Dudock. New York: Academic Press, 1987.

1969

☆

Physics
MURRAY GELL-MANN
for his contributions and discoveries concerning the classification
of elementary particles and their interactions.

☆

Chemistry
SIR DEREK H. R. BARTON and **ODD HASSEL**
for their contributions to the development of the concept
of conformation and its application in chemistry.

☆

Physiology or Medicine
MAX DELBRÜCK, ALFRED D. HERSHEY
and **SALVADOR E. LURIA**
for their discoveries concerning the replication
mechanism and the genetic structure of viruses.

☆

Literature
SAMUEL BECKETT
for his writing, which - in new forms for the novel and drama
- in the destitution of modern man acquires its elevation.

☆

Peace
INTERNATIONAL LABOUR ORGANIZATION (I.L.O.) Geneva.

☆

Economics
RAGNAR FRISCH and **JAN TINBERGEN**
for having developed and applied dynamic models
for the analysis of economic processes.

1969

Max Ludwig Henning Delbrück (1906-1981); Salvador Edward Luria (1912-1991); and Alfred Day Hershey (1908-1997).

Max Ludwig Henning Delbrück

Salvador Edward Luria

Alfred Day Hershey

© The Nobel Foundation

Max Delbrück (left), **Salvador Luria** (right), and **Alfred Hershey,** won the Nobel Prize in Physiology or Medicine for their discoveries concerning "the replication mechanism and the genetic structure of viruses."

Son of a history professor in Berlin, Delbrück earned his Ph.D., in physics from the University of Göttingen in 1930. A celebrated lecture by Niels Bohr in 1932, "Light and Life," influenced Delbrück so much that he formed an abiding interest in the study of biology. Using principles of physics, he tried to find how genes maintain long-term stability yet undergo short-term mutations. While Delbrück was visiting the United States on a fellowship in 1937, World War II broke out, preventing his return to Europe. In 1940 he joined Vanderbilt University in Nashville, Tennessee, and started working with Salvador

Luria, an Italian immigrant who had been a Guggenheim fellow at that institution. Luria was also interested in adapting the principles of physics and chemistry for biological research.

Using the new technology of the electron microscope, Delbrück and Luria determined that spontaneous genetic mutations were responsible for the ability of some bacteria to resist the destructive actions of the bacteriophage. Their work attracted Alfred Hershey, a bacteriologist trained at Michigan State College (now Michigan State University), who had been working on the genetics of bacteriophage viruses at Washington University in St. Louis. Soon, the trio began one of the most fruitful collaborations in the history of virology and genetics.

In 1946, Hershey demonstrated that a single virus may contain more than one gene explaining the independence of different mutations. Delbrück and Hershey also discovered that bacteriophages (the viruses that 'infect' bacteria) exchanged genetic materials such that the offspring had different infective capabilities than either of their parent-viruses. In 1952, one year before Watson, Crick and Wilkins discovered the structure of the DNA molecule, Hershey and his colleague Martha Chase proved that genes were in fact, made up of DNA.

In his boyhood, Delbrück lived in an affluent neighborhood of Berlin. It seems that he would steal cherries from the yard of the famous physicist Max Plank. Delbrück grew in a highly intellectual milieu at a time of great political and social upheaval. An intense humanist, he constantly searched for simple, unifying truth that explained the relationship between the world of living organisms and the "incorruptible" physical universe.

In April 1943, upon Hershey's invitation, Delbrück and Luria traveled to St. Louis to discuss research, a meeting Hershey would later call the "first phage meeting." In 1945, Delbrück taught the first 'phage course' at the Cold Spring Harbor Laboratory in New York. Later, he and his collaborators remained active in the "phage group" — an informal group of scientists, dedicated to free exchange of ideas on genetics, biology, and life. With exchange of their research findings and data, they avoided duplicate efforts and undue competition, setting up exemplary ethical standards in research.

It seems that about a year after their marriage in 1941, Max Delbrück's wife, Mary Bruce Delbrück wrote a delightful account of her husband's scientific-minded marital life and reported discovering a "new species" to which her husband belonged—the title of her article was *Homo Scientificus*.

References

1. Delbrück M. Radiation and the Hereditary Mechanism. The American Naturalist, vol. 74, p. 350-362, 1940
2. Delbrück M. The Growth of Bacteriophage and Lysis of the Host.
3. Delbrück M. Cosmic Rays and the Origin of Species. Nature, vol. 137, pp. 358-359, 1936.
4. Delbrück M. Mind and Matter? In The Nature of Life, edited by William H. Heidcamp, 1978.
5. Hershey AD (with Luria S). A Slot Machine, a Broken Test Tube. New York: Harper & Row, 1984.
6. Hershey AD (with Rotman R). Genetic Recombination Between Host-Range and Plaque-Type. Genetics, vol. 34, 1949, pp. 44-71.
7. Hershey AD. Spontaneous Mutations in Bacterial Viruses. Genetics, vol. 11, 1946, pp. 67-76.
8. Luria SE (with Delbrück). Mutations of Bacteria from Virus Sensitivity to Virus Resistance. Genetics, vol. 28, pp. 491-511, 1943.
9. Luria SE. Genetics of Bacteriophage. Annual Review of Microbiology, vol. 26: 205-240, 1962.

1970

Physics
HANNES ALFVÉN
*for fundamental work and discoveries in magneto-hydrodynamics
with fruitful applications in different parts of plasma physics, and*
LOUIS NÉEL
*for fundamental work and discoveries concerning anti-ferromagnetism and
ferrimagnetism which have led to important applications in solid state physics.*

Chemistry
LUIS F. LELOIR
*for his discovery of sugar nucleotides and their role
in the biosynthesis of carbohydrates.*

Physiology or Medicine
SIR BERNARD KATZ, ULF VON EULER and JULIUS AXELROD
*for their discoveries concerning the humoral transmitters in the nerve
terminals and the mechanism for their storage, release and inactivation.*

Literature
ALEKSANDR ISAEVICH SOLZHENITSYN
*for the ethical force with which he has pursued the
indispensable traditions of Russian literature.*

Peace
NORMAN BORLAUG
Led research at the International Maize and Wheat
Improvement Center, Mexico City.

Economics
PAUL A SAMUELSON
*for the scientific work through which he has developed static
and dynamic economic theory and actively contributed to
raising the level of analysis in economic science.*

1970

| Sir Bernard Katz | Ulf Svante von Euler | Julius Axelrod |

© The Nobel Foundation

Bernard Katz (left), **Ulf von Euler** (right), **and Julius Axelrod** Three scientists shared the 1970 Nobel Prize in Physiology or Medicine for their discoveries concerning "the humoral transmitters in the nerve terminals and the mechanisms for their storage, release and inactivation."

Born in Leipzig, Katz studied medicine there and obtained a doctorate in physiology from the University of London. He served briefly in the Royal Australian Air Force during World War II before returning to the faculty of London's University College. Focusing their interest on neurotransmitter research, Katz and colleagues used minute, electrolyte-filled glass pipettes with 0.5-micron tips and recorded motor end-plate potentials during acetylcholine-mediated synaptic transmissions. They found that small amounts of acetylcholine were released constantly from the terminal vesicles of the neurons, disproving the prevalent view that during rest the neurons transmitted no signals. In 1977, Katz also discovered that the release of acetylcholine was due to the movement of the ionic calcium ions across neuronal gaps.

Axelrod was born in New York and studied there. In 1949, he joined the National Institutes of Health as a research scientist. Focusing his work on sympathomimetic amines and their metabolism, Axelrod discovered (and named) catechol-ø-methyltransferase (COMT) enzyme that catabolized norepinephrine. He also discovered a new class of microsomal enzymes in hepatic cells, later named cytochrome P-450 monooxygenases. His work on norepinephrine release, activation, storage and reuptake in the brain became central to understanding the mechanism of action of neuro-pharmacological agents, and to the development of drugs for many psychotic disorders. Axelrod also discovered the role of the pineal gland in the synthesis of melatonin.

Von Euler's father and father's godfather were both Nobel laureates in chemistry. His father Hans von Euler-Chelpin had won the 1929 award for work on sugar fermentation, and his fathers' godfather, Svante Arrhenius had won his award in 1903 for work on the electrolytic theory of dissociation.

Ulf von Euler was born in Stockholm and obtained his medical degree at the Karolinska Institute, where he also worked much of his life. Von Euler discovered the neurotransmitter norepinephrine and characterized its pharmacology and metabolism. Respected as the one of the rare intellectual giants from the classical era of physiology and pharmacology, Von Euler made two other significant discoveries early in his career. Working with Henry Dale in London, he isolated 'substance P' in 1931. In 1935, he discovered and isolated a new chemical from the prostate glands of sheep and humans, which he named "prostaglandin."

References

1. Alexrod J (with Brodie BB). The Fate of Acetophenitidin (Phenacetin) in Man and a method for the Estimation of Acetophenitidin and its Metabolites in Biological Material. Journal of Pharmacology and Experimental Therapeutics, vol. 97, pp. 58-67, 1949.

2. Alexrod J (with Whitby and Hertling). Effect of Cocaine on the Disposition of Noradrenaline Labeled with Tritium. Nature, vol. 187, pp. 604-605, 1960.

3. Katz B (with Del Castillo J). Electrical Excitation of the Nerve. Journal of Physiology; vol. 124, pp. 553-559, 1954.

4. Katz B (with Del Castillo J). Statistical Factors Involved in Neuro-Muscular Facilitation and Depression. Journal of Physiology, vol. 124, pp. 574-585, 1954.

5. Katz B (with Moore, J). Membranes, Ions, and Impulses. New York: Plenum, 1975.

6. Von Euler US. Noradrenaline: Chemistry, Physiology, Pharmacology, and Clinical Aspects, 1956.

1971

☆

Physics

DENNIS GABOR

for his invention and development of the holographic method.

☆

Chemistry

GERHARD HERZBERG

*for his contributions to the knowledge of electronic structure
and geometry of molecules, particularly free radicals.*

☆

Physiology or Medicine

EARL W. JR. SUTHERLAND

for his discoveries concerning the mechanisms of the action of hormones.

☆

Literature

PABLO NERUDA

*for a poetry that with the action of an elemental force
brings alive a continent's destiny and dreams.*

☆

Peace

WILLY BRANDT

Federal Republic of Germany, Chancellor of the Federal Republic
of Germany, initiator of West Germany's *Ostpolitik*.

☆

Economics

SIMON KUZNETS

*for his empirically founded interpretation of economic growth
which has led to new and deepened insight into the economic
and social structure and process of development.*

☆

Peace
WILLY BRANDT
Federal Republic of Germany, Chancellor of the Federal Republic of Germany, initiator of West Germany's *Ostpolitik.*

☆

Economics
SIMON KUZNETS
for his empirically founded interpretation of economic growth which has led to new and deepened insight into the economic and social structure and process of development.

1971

EARL WILBUR SUTHERLAND, JR. (1915-1974)

© The Nobel Foundation

Earl Sutherland was awarded the 1971 Nobel Prize in Physiology or Medicine for the discovery of cyclic AMP and demonstration of its role as a "second messenger."

He was born in Berlingame, Kansas. When he was in high school, Sutherland was deeply inspired after reading de Kruif's *Microbe Hunters* (Harcourt, Brace, Jovanovich, 1926) and took up medicine as a career. After obtaining his MD from Washington University in St. Louis, he served in the United States Army for two years before rejoining his alma mater as a lecturer in pharmacology. In 1953, he moved to Case Western Reserve University in Cleveland.

Carl and Gerty Cori's laboratory at Washington University was an ideal place for enzyme research. There, Sutherland explored the biochemistry of epinephrine action. Using liver slices, he showed that epinephrine and glucagon-initiated glycogen breakdown by activating liver phosphorylase, a glycolytic enzyme discovered by the Coris. With his colleague Ted Rall, Sutherland then established that epinephrine and glucagon helped maintain a critical balance between activation and deactivation of phosphorylase, and hence the amount of glucose breakdown from the liver.

Sutherland and coworkers used debris from liver cell homogenates after centrifugation, and found a heat-stable, dialyzable factor involved in epinephrine and glucagon action. They named the chemical "adenine ribonucleotide," later identified as 3', 5' adenosine monophosphate, or cyclic AMP (cAMP). Sutherland discovered that besides in liver, the compound was abundant in cell debris from the heart, skeletal muscles and the brain. In 1962, he isolated adenyl cyclase, an enzyme associated with cell membranes, which converted ATP into cAMP.

The discovery of cAMP was a major step in understanding the molecular basis of hormone functions and drug actions upon the immune system, brain, heart, and the smooth muscle. A superb clinician and a man of exceptional vision, Sutherland referred to hormones as the "first messengers" stimulating the formation of cAMP, which he called the "second messenger." Sadly, within three years of receiving the Nobel Prize, Sutherland died of a massive esophageal hemorrhage.

References

1. Sutherland EW (with Robison and Butcher). Cyclic AMP. 1971.
2. Sutherland EW. Formation of a Cyclic Adenine Ribonucleotide by Tissue Particles. Journal of Biological Chemistry, vol. 232, pp. 1065-1076, 1958.
3. Sutherland EW. The Biological Role of Adenosine 3', 5' Phosphate," Harvey Lecture Series, vol. 57, pp. 17-33, 1962.
4. Sutherland EW. The Role of the Cyclic 3', 5'-AMP in Response to Catecholamines and Other Hormones." Pharmacological Review, vol. 18, pp. 145-161, 1966.

1972

☆

Physics
JOHN BARDEEN, LEON N. COOPER
and **J. ROBERT SCHRIEFFER**
for their jointly developed theory of superconductivity,
usually called the BCS-theory.

☆

Chemistry
CHRISTIAN B. ANFINSEN
for his work on ribonuclease, especially concerning the connection between
the amino acid sequence and the biologically active confirmation
and
STANFORD MOORE and **WILLIAM H. STEIN**
for their contribution to the understanding of the connection between chemical
structure and catalytic activity of the active center of the ribonuclease molecule.

☆

Physiology or Medicine
GERALD M. EDELMAN and **RODNEY R. PORTER**
for their discoveries concerning the chemical structure of antibodies.

☆

Literature
HEINRICH BÖLL
for his writing which through its combination of a broad
perspective on his time and a sensitive skill in characterization
has contributed to a renewal of German literature.

☆

Peace
No Award

☆

Economics
SIR JOHN R. HICKS and **KENNETH J. ARROW**
for their pioneering contributions to general economic
equilibrium theory and welfare theory.

1972

GERALD MAURICE EDLEMAN (1929-2014)
AND RODNEY R PORTER (1917-1985)

Gerald Maurice Edelman   **Rodney Robert Porter**

© The Nobel Foundation

Gerald Edelman (left), and **Rodney Porter** received the 1972 Nobel Prize in Physiology or Medicine for their discoveries concerning the chemical structure of antibodies.

Porter was born in Newton-le-Willows, Lancashire, UK, and studied in Liverpool and Cambridge. In 1960, he became the first Pfizer Professor of immunology at St. Mary's Hospital and in 1967, he went to Oxford as Whitley Professor of Biochemistry. While at Cambridge, Frederick Sanger, the 1956 Nobel laureate in Chemistry, inspired Porter to study the structure of proteins. Using the papain enzyme, Porter tried breaking the large immunoglobulin molecule (IgG) and obtained a single crystalizable piece (Fc), devoid of antibody binding, and two fragments that retained antibody-binding properties (Fab).

Meanwhile, in the mid1960s, Gerald Edelman, a physician trained in biochemistry at Rockefeller University had shown that an immunoglobulin G molecule can be broken into two pairs of "chains" held together by disulfide bonds. He called them heavy and light chains based on their molecular weights. In a series of "antibody workshops" organized world over, Porter and Edelman along with other immunologists began sharing ideas and data.

Porter then showed that light and heavy chains of Edelman reacted with the Fab antiserum, while only the heavy chains reacted with the Fc antiserum. He proposed that the structure of the antibody molecule resembled the letter Y. The branches of the Y consisted of one light chain and one half of the heavy chain (Fab), and the Y's stem of the other half of the heavy chain (Fc). Later, Porter elucidated the amino acid sequence of the heavy chain and studied the structure and functions of many proteins from the complement system. Unfortunately, at the peak of his fame, porter was killed in an automobile accident in 1985.

By 1969, Edelman and colleagues worked out the entire sequence of 1,300 amino acids of immunoglobulin G—then the largest molecule to be biochemically defined. In 1982, Edelman became the director of Scripps Research Institute, La Jolla, California. In his books *Neural Darwinism* (1987) and *Bright Air, Brilliant Fire* [(1992) both by Basic Books, Inc.], he explained with great charm, the neural basis of consciousness. He proposed the "neuronal-group selection theory"—a melding of molecular biological concepts with that of Darwinian natural selection. A multifaceted intellectual giant, Edelman is an avid sportsman, poet, and an accomplished violinist.

References

1. Edelman GM (with Cunningh B, Gall WE, Gottlieb PD, Rutishauser U, and Waxddal MJ). The Covalent Structure of the Entire γG Immunoglobulin Molecule. Proceedings of the National Academy of Sciences, vol. 63, pp. 78-85, 1969.

2. Edelman GM. Dissociation of γGlobulin. Journal of the American Chemical Society. Vol. 81, pp. 3155-3156, 1959.

3. Edelman GM. Neural Darwinism: The Theory of Neuronal Group Selection, 1987.

4. Porter RR. The Hydrolysis of Rabbit γ-Globulin and Antibodies with Crystalline Papain," Biochemical Journal, vol. 73, 1959. p. 119. Immunochemistry, Annual Review of Biochemistry, vol. 31, 1962, pp. 625-647; Structure and Activation of the Early Components of Complement, Federation Proceedings, vol. 36, 1977, pp. 2191-2196.

5. Porter RR. The Major Histocompatibility Complex and Diseases. Molecular Biology and Medicine, vol. 1, pp. 161-168, 1983. Complement and Autoimmunity. Bioassay, vol. 1, pp. 261-264, 198

1973

☆

Physics
LEO ESAKI and IVAR GIAEVER
for their experimental discoveries regarding tunneling phenomena in semiconductors and superconductors, respectively, and
BRIAN D. JOSEPHSON
*for his theoretical predictions of the properties of a super-current through a tunnel barrier, in particular those phenomena which are generally known as the "**Josephson** effects."*

☆

Chemistry
ERNST OTTO FISCHER and SIR GEOFFREY WILKINSON
for their pioneering work, performed independently, on the chemistry of the organometallic, so called sandwich compounds.

☆

Physiology or Medicine
KARL VON FRISCH, KONRAD LORENZ and NIKOLAAS TINBERGEN
for their discoveries concerning organization and elicitation of individual and social behavior patterns.

☆

Literature
PATRICK WHITE
for an epic and psychological narrative art which has introduced a new continent into literature.

☆

Peace
HENRY A. KISSINGER
Secretary of State, State Department, Washington. And
LE DUC THO , Democratic Republic of Viet Nam. (Declined the
prize.) *for jointly negotiating the Vietnam peace accord in 1973.*

☆

Economics
WASSILY LEONTIEF

for the development of the input-output method and for its application to important economic problems.

1973

KARL VON FRISCH (1886-1982),
KONRAD LORENZ, (1903-1989) AND
NIKOLAAS TINBERGEN (1907-1988)

Karl Ritter von Frisch   **Konrad Zacharias Lorenz**   **Nikolaas Tinbergen**

© The Nobel Foundation

Karl von Frisch (left), **Konrad Lorenz** (right), and **Nikolaas Tinbergen** shared the 1973 Nobel Prize in Medicine or Physiology "for their discoveries concerning organization and elucidation of individual and social behavior patterns."

Son of a surgeon, Frisch was born in Vienna. His uncle, Sigmund Exner, instilled in him a love for animals and a desire to study zoology. Frisch spent many summers in Brunnwinkl, a rustic Austrian village, where an old water mill had been converted into a summer resort. It was here that he did most of his famous studies on honeybees.

Biologists then believed that bees were color-blind. Frisch tested this in simple studies. Using dishes of sugar water on colored papers, he showed that bees were indeed, not color-blind; in fact, they can be 'taught' to see dishes with sugar water on colored papers. Over the next

15 years, Frisch studied the 'language' of honeybees. He showed that bees danced in circular, straight, figure-eight, or tail-wagging patterns to convey the precise distances and locations of the forage in relation to the sun.

The Vienna-born biologist Lorenz also loved animals since childhood; he studied the social behavior of chicks, ducklings and other young animals. Minutes after the egg was hatched, he found that ducklings and goslings can be made to follow any moving object (including himself as he swam in a pond), as if the object were their mother. By this "imprinting" or programmed learning, the adult passes on survival skills to its young with supreme efficiency. Later, Lorenz studied human aggression. He wrote many books for the general reader, in which he explored the relevance for human societies, of learning about animal behavior.

In a career spanning over five decades, the Dutch zoologist Tinbergen also studied animals ranging from digger wasps to sled dogs; from long-necked gulls to winged butterfly; and from giant caterpillars to colored moths. Using methods that are now classic, he tested animal in their habitats and explained the nature of 'instinct' and survival strategies. Tinbergen's thirty-two-page thesis was the shortest ever at Leiden University to be awarded a Ph.D. A modest, soft-spoken man, it seems that he felt embarrassed to have won the Nobel Prize. The contributions of these three scientists led to the growth of the new discipline of neuroethology—the neurobiology of behavior.

References

1. Lorenz K (with Cloud W) Winners and Sinners. The Sciences 13, pp. 16-21, December 1973.
2. Lorenz K (with Gould JL) The Mechanisms and Evolution of Behavior. New York: W.W. Norton, 1982.
3. Lorenz K (with Lehrman DS). A Critique of Konrad Lorenz's Theory of Instinctive Behavior. Quarterly Review of Biology 28, pp. 337-363, 1953.

4. Lorenz K. Evolution and Modification of Behavior, 1965.
5. Lorenz K. King Solomon's Ring, 1952.
6. Tinbergen N. An Objectivistic Study of the Inmate Behavior of Animals, 1942.
7. Tinbergen N. Curious Naturalists. New York: Basic Books, 1958.
8. Tinbergen N. The Animal in its World, vol. 1, 1972.
9. Tinbergen N. The Tale of John Stickle, 1954.
10. Von Frisch K (with Eibl-Eibesfeldt I). The Biology of Behavior. New York: Holt, Reinhart and Winston, 1970.
11. Von Frisch K (with Lindauer M). Communication Among Social Bees. Cambridge, Mass.: Harvard University Press, 1961.
12. Von Frisch K. Decoding the Language of the Bee. Science, vol. 185, pp. 663-669, 1974.
13. Von Frisch K. Man and the Living World. Biology, 1964.

1974

☆

Physics
SIR MARTIN RYLE and ANTONY HEWISH
for their pioneering research in radio astrophysics Ryle for his observations and inventions, in particular of the aperture synthesis technique, and Hewish for his decisive role in the discovery of pulsars.

☆

Chemistry
PAUL J. FLORY
for his fundamental achievements, both theoretical and experimental, in the physical chemistry of the macromolecules.

☆

Physiology or Medicine
ALBERT CLAUDE, CHRISTIAN DE DUVE and GEORGE E. PALADE
for their discoveries concerning the structural and functional organization of the cell.

☆

Literature
EYVIND JOHNSON
for a narrative art, farseeing in lands and ages, in the service of freedom, and
HARRY MARTINSON
for writings that catch the dewdrop and reflect the cosmos.

☆

Peace
SEÁN MAC BRIDE
President of the International Peace Bureau, Geneva and the Commission of Namibia, United Nations, New York, and
Eisaku Sato
Prime Minister of Japan.

☆

Economics
GUNNAR MYRDAL and FRIEDRICH AUGUST VON HAYEK

for their pioneering work in the theory of money and economic fluctuations and for their penetrating analysis of the interdependence of economic, social and institutional phenomena.

1974

ALBERT CLAUDE (1899-1983), GEORGE
EMIL PALADE (1912-2008), AND
CHRISTIAN RENÉ DE DUVE (1917-2013)

Albert Claude   George Emil Palade   Christian René
de Duve

© The Nobel Foundation

Albert Claude (left)**, George Emil Palade** (right)**,** and **Christian de Duve** shared the 1974 Nobel Prize for physiology or medicine "for their discoveries concerning the structural and functional organization of the cell."

Albert Claude was born in Longlier, Belgium. Financial difficulties forced him to end schooling at age 12. During World War I, he served in the British underground, for which he earned a valor citation from Winston Churchill. Claude's wartime services also enabled him to enroll in college without graduating from high school. After earning a medical degree from the University of Liège, he sailed to the United States in 1929, where he worked for 20 years at the Rockefeller Institute.

Claude developed a high-speed differential centrifugation method, using which he tried isolating cancer-causing agents from cell-free filtrates of tumors. Although this effort failed, he discovered numerous 'granular' materials in the normal tissues used as controls. These, he thought, might be the mitochondria. In a profound understatement, he predicted that isolation of mitochondria "should lead to interesting developments."

George Palade was born in Jassy, a town in Eastern Romania. He came to the United States for advanced studies and joined Claude in 1948. Using the differential centrifugation method and the electron microscope "to the highest degree of artistry," Palade and Claude fractionated cell organelles and discovered ribosome and the endoplasmic reticulum (ER). They showed that the ribosomes were rich in RNA and protein complexes and played a major role in protein synthesis. They also found that protein transportation occurred via the ER to the Golgi apparatus, and that it was in the mitochondria where energy metabolism took place.

De Duve's parents fled Belgium during World War I and lived near London when Christian was born. The family returned to Belgium in 1920. After education in humanities and literature, de Duve studied medicine at the Catholic University of Louvain. Using the Palade technique of centrifugation, he separated cellular organelles according to their size, shape and density, maintaining their function; he discovered the mysterious intracellular structure, lysosomes, and showed that they were the intracellular 'suicide bags' involved in catalysis of chemical molecules. He later discovered peroxisomes (other intracellular organelle) and established their role in fatty acid metabolism. These findings were to shed light on the pathogenesis of several inherited metabolic disorders.

The discoveries made and the research methods developed by these scientists had a major impact in founding a new branch of science—cell and molecular biology.

References

1. de Duve C (with Berthet J). Glucose, insulin, et diabetes, 1945; Tissue Fractionation Studies: 1, The Existence of a Mitochondria-

Linked Enzymatically Inactive Form of Acid Phosphatase in Rat-Liver Tissue. Biochemical Journal, vol. 50, pp. 174-181, 1951.

2. de Duve C (with Dean RT). Lysosomes. London: Edward Arnold, 1977.

3. Palade GE. Cytochemical Studies of Mammalian Tissues: I, Isolation of Intact Mitochondria from Rat Liver, Some Biochemical Properties of Mitochondria and Submicroscopic Particulate Material. Journal of Biological Chemistry, vol. 172, pp. 619-635, 1948.

4. Palade GE. Studies on the Endoplasmic Reticulum: I, Its Identification in Cells in Situ. Journal of Experimental Medicine, vol. 100, pp. 641-656, 1954.

5. Palade GE. The Fine Structure of Mitochondria. Anatomical Record, vol. 114, pp. 427-451, 1952.

1975

☆

Physics
AAGE BOHR, BEN MOTTELSON and JAMES RAINWATER
for the discovery of the connection between collective motion and particle motion in atomic nuclei and the development of the theory of the structure of the atomic nucleus based on this connection.

☆

Chemistry
SIR JOHN WARCUP CORNFORTH
for his work on the stereochemistry of enzyme-catalyzed reaction, and
VLADIMIR PRELOG
for his research into the stereochemistry of organic molecules and reactions.

☆

Physiology or Medicine
DAVID BALTIMORE, RENATO DULBECCO and HOWARD MARTIN TEMIN
for their discoveries concerning the interaction between tumor viruses and the genetic material of the cell.

☆

Literature
EUGENIO MONTALE
for his distinctive poetry which, with great artistic sensitivity, has interpreted human values under the sign of an outlook on life with no illusions.

☆

Peace
ANDREI DMITRIEVICH SAKHAROV
Soviet nuclear physicist. Campaigner for human rights.

☆

Economics
LEONID VITALIYEVICH KANTOROVICH and TJALLING C. KOOPMANS
for their contributions to the theory of optimum allocation of resources.

1975

RENATO DULBECCO (1914-2012),
DAVID BALTIMORE (BORN 1938), AND
HOWARD MARTIN TEMIN (1934-1994)

Renato Dulbecco   David Baltimore   Howard Martin Temin

© The Nobel Foundation

Renato Dulbecco (left), **David Baltimore** (right), and **Howard Martin Temin** shared the Nobel Prize in Physiology or Medicine for their "discoveries concerning the interaction between tumor viruses and the genetic material of the cell."

Dulbecco was born in Catanzaro, Italy, and studied medicine in Turin. He served as military officer in the Italian Army, but after German occupation, he joined the Resistance during World War II. In 1947, he began working with Salvador Luria at the Indiana University in Bloomington on the genetics of bacteriophage.

Dulbecco applied his expertise from bacteriophage research to the study of tumor viruses at the California Institute of Technology (Caltech) with Max Delbrück. Later he collaborated with David Baltimore at the Salk Institute. Baltimore, a New Yorker trained at the Massachusetts Institute of Technology, was a research associate at the Salk Institute. These scientists discovered that the invading virus carried an enzyme that changed the DNA of the infected cell. Dulbecco also developed the plaque-assay technique for measuring the viral load in cell cultures—a method that helped in developing the polio vaccine.

Born in Philadelphia and educated at Caltech, Howard Temin worked most of his career in the University of Wisconsin's Oncology Department. With Harry Rubin, a postdoctoral fellow, he studied the Rous sarcoma virus, focusing on the RNA-mediated DNA synthesis using improved viral culture techniques. In 1965 Temin proposed the phenomenon of reverse transcription in which the viral RNA influences the host DNA by inserting its own genes.

Initially, this proposal was not accepted because the prevailing dogma in genetics was that DNA transmitted genetic information through RNA—not vice versa. By 1970, using Rauscher mouse leukemia virus, Baltimore discovered an RNA viral enzyme that copied the genetic information to the host DNA. Around the same time, Temin independently discovered a similar enzyme in the Rous sarcoma virus; the enzyme later became known as *reverse transcriptase.*

Although not fully appreciated at the time, the discovery of reverse transcriptase helped understand how normal cells become cancerous—the knowledge that influenced research in such diverse fields as oncology, virology, genetic engineering, and AIDS.

References

1. Baltimore D (with Huang AS). Isopycnic Separation of Subcellular Components from Poliovirus-Infected and Normal HeLa Cells. Science, vol. 162, pp. 572-574, 1968.
2. Baltimore D An RNA-Dependent DNA Polymerase in Virions of RNA Tumor Viruses. Nature, vol. 226, pp. 1209-1211, 1970.
3. Baltimore D. RNA-Directed DNA Synthesis and RNA Tumor Viruses. Advances in Virus Research, vol. 17, pp. 51-94, 1972.
4. Barry RD, and Mahy BWJ. The Biology of Large RNA Viruses. New York: Academic Press, 1970.

5. Dulbecco R (with Luria SE). Reactivation of Ultraviolet-Inactivated Bacteriophage by Visible Light. Nature, vol. 163, pp. 949-950, 1949.

6. Dulbecco R. The Design of Life. New Haven, Conn.: Yale University Press, 1987.

7. Dulbecco R. Genetic Re-combinations Leading to Production of Active Bacteriophage from Ultraviolet Inactivated Bacteriophage Particles. Genetics, vol. 34, pp. 93-125, 1949.

8. Fraenkel-Conrat H. Design and Function at the Threshold of Life: The Viruses. New York: Academic Press, 1962.

9. Temin HM. Nature of the Provirus of Rous Sarcoma. National Cancer Institute Monograph, vol. 17, pp. 557-570, 1964.

10. Temin HM. The Effects of Actinomycin D on Growth of Rous Sarcoma Virus in Vitro. Virology, vol. 20, pp. 577-582, 1963.

11. Temin HM. The Participation of DNA in Rous Sarcoma Virus Production. Virology, vol. 23, pp. 486-494, 1964.

1976

☆

Physics
BURTON RICHTER and SAMUEL C. C. TING

*for their pioneering work in the discovery of a heavy
elementary particle of a new kind.*

☆

Chemistry
WILLIAM N. LIPSCOMB

*for his studies on the structure of boranes illuminating
problems of chemical bonding.*

☆

Physiology or Medicine
BARUCH S. BLUMBERG and D. CARLETON GAJDUSEK

*for their discoveries concerning new mechanisms for the
origin and dissemination of infectious diseases.*

☆

Literature
SAUL BELLOW

*for the human understanding and subtle analysis of
contemporary culture that are combined in his work.*

☆

Peace
BETTY WILLIAMS and MAIREAD CORRIGAN

Founders of the Northern Ireland Peace Movement
(later renamed Community of Peace People).

☆

Economics
MILTON FRIEDMAN

*for his achievements in the fields of consumption analysis, monetary history
and theory and for his demonstration of the complexity of stabilization policy.*

1976

BARUCH SAMUEL BLUMBERG (1925-
2011), AND DANIEL CARLETON
GAJDUSEK (1923-2011)

Baruch Samuel Blumberg   **Daniel Carleton Gajdusek**

© The Nobel Foundation

Baruch S. Blumberg (left), and **Daniel Carleton Gajdusek** received the 1976 Nobel Prize for "discoveries concerning new mechanisms for the origin and dissemination of infectious diseases."

Blumberg was born in New York and graduated in medicine from Columbia University. After earning a Ph.D., in biochemistry from the University of Oxford, he headed the geographic medicine division at the National Institutes of Health (NIH) in Bethesda.

Blumberg was interested in polymorphism of proteins and hemoglobin in different ethnic groups. Using agar gel electrophoresis, he and coworkers searched for antibodies in patients who had received multiple blood transfusions. In 1963, they isolated a protein from a hemophilic patient, which reacted with the serum from an Australian aborigine; hence, they named the protein Australia antigen (Au). In 1967, Barbara Werner, a technician in Blumberg's laboratory, contracted hepatitis and her blood became positive for Au, suggesting a link between hepatitis and Au. Soon, Au screening of all blood donors began, and only Au-negative blood was deemed safe for transfusion. This practice dramatically reduced post-transfusion hepatitis. In the 1970s, hepatitis B virus was identified, and Au was shown to be the surface antigen of the virus. Blumberg also pioneered research on developing a vaccine for hepatitis B.

Gajdusek was born in Yonkers, New York, and graduated in medicine from Harvard Medical School, specializing in pediatric neurology. In 1958, he joined the National Institutes of Neurological Disorders and Stroke at the NIH.

During his postdoctoral studies in Melbourne, Gajdusek visited New Guinea and saw victims of kuru, a fatal neurological disease among the Fore aborigines. Gajdusek designed years of fieldwork. With his knowledge of anthropology, child development, and neurology, he learned the culture of the Fore people, understood the epidemiology of kuru and succeeded in transmitting the disease to laboratory baboons. He proposed that kuru was a 'slow virus' infection propagated through ritual cannibalism, in which women and children ate brains of the dead. Eventually, educational efforts all but eradicated kuru in New Guinea.

With recent interest in bovine spongiform encephalopathy and diseases caused by prion, Gajdusek's work has acquired new meaning. A brilliant scientist, Gajdusek has written extensively on diverse subjects, including anthropology and the people of New Guinea.

References

1. Blumberg BS (with Sutnic AI and London WT). Australia Antigen: A Genetic Basis for Chronic Liver Disease and Hepatoma? Annals of Internal Medicine, vol. 74, pp. 442-443, 1971.

2. Blumberg BS. Bioethical Questions Related to Hepatitis B Antigen. American Journal of Clinical Pathology, vol. 65, pp. 848-853, 1976.

3. Blumberg BS. The Hepatitis B Virus: Landmarks in American Epidemiology. Public Health Reports, vol. 95, pp. 427-435, 1980.

4. Bumberg BS. Hepatitis: The Plight of the Carrier. The Sciences March, vol. 30, pp. 10-11, 1978.

5. Eron C. The Virus That Ate Cannibals. New York: MacMillan, 1981.

6. Gajdusek DC (with Gibbs CJ, Asher DM, and Alpers MP). Transmission to the Chimpanzee. Science, vol. 161, pp. 388-389, 1968.

7. Gajdusek DC (with Zigas V). Kuru: Clinical Study of a New Syndrome Resembling Paralysis Agitans in Natives of the Eastern Highlands of Australian New Guinea. The Medical Journal of Australia, vol. 2, pp. 745-754, 1957.

1977

☆

Physics

ILYA PRIGOGINE

for his contributions to non-equilibrium thermodynamics,
particularly the theory of dissipative structures.

☆

Chemistry

PHILIP W. ANDERSON, SIR NEVILL F.
MOTT and JOHN H. VAN VLECK

for their fundamental theoretical investigations of the
electronic structure of magnetic and disordered systems.

☆

Physiology or Medicine

ROGER GUILLEMIN and ANDREW V. SCHALLY

for their discoveries concerning the peptide
hormone production of the brain, and

ROSALYN YALOW

for the development of radio-immunoassays of peptide hormones.

☆

Literature

VICENTE ALEIXANDRE

for a creative poetic writing which illuminates man's condition in the
cosmos and in present-day society, at the same time representing the
great renewal of the traditions of Spanish poetry between the wars.

☆

Peace

AMNESTY INTERNATIONAL

London, Great Britain. A worldwide organization for the
protection of the rights of prisoners of conscience.

☆

Economics

BERTIL OHLIN and JAMES E MEADE

for their path-breaking contribution to the theory of
international trade and international capital movements.

1977

ROGER CHARLES LOUIS GUILLEMIN (1924-2024), ANDREW VICTOR SCHALLY (1926-2024) AND ROSALYN SUSSMAN YALOW (1921-2011)

Roger Charles Louis Guillemin

Andrew Viktor Schally

Rosalyn Sussman Yalow

© The Nobel Foundation

The 1977 Nobel Prize was awarded to **Roger Charles Louis Guillemin** (left) **and Andrew Victor Schally** (right) "for their discoveries of the peptide hormone production of the brain," and to **Rosalyn Yalow** "for the development of radioimmunoassay."

Born in Dijon, France, Roger Guillemin studied there and in Montreal, joining the Baylor College of Medicine in Houston in 1953. Geoffrey W. Harris, a British scientist showed that hypothalamus controlled pituitary secretions through the capillary system connecting them. This spurred Guillemin and others to begin an intense search for the hypothalamic chemicals. Guillemin obtained 5 million pieces of sheep hypothalamic from slaughterhouses in France, and by 1968, isolated 1 mg of thyrotropin releasing factor (TRF). With colleague Richard Burgus, Guillemin then isolated luteinizing hormone-releasing factor (LRF) and somatostatin. His team also worked out the structure of hypothalamic hormones.

In 1954, Andrew Schally, a Polish biochemist trained in Montreal, joined Guillemin at Baylor; but their relationship was not cordial. In 1962, Schally moved to the Veterans Administration Hospital in New Orleans, and continued isolation of hypothalamic hormones from the pig thalami. Almost at the same time as Guillemin, Schally too, isolated TRH, LRH and somatostatin, and worked out the structures of many of the hormones. An avid soccer player, Schally is also famous for his brilliance and hard work.

Yalow was born in New York and studied physics at the University of Illinois. In 1950, she joined Dr. Solomon Berson at the New York University College of Medicine. Her knowledge of physics and Berson's clinical expertise helped the duo develop radio-immune assay (RIA) method, with which they discovered insulin antibodies. The simple, elegant and sensitive Yalow-Berson RIA became the basic tool to measure minute quantities of hormones, enzymes, and proteins, in clinical care and in advanced biomedical research.

Columbia University denied Yalow's postgraduate entry into physics because she was a woman. She was asked to learn stenography and work as secretary to attend physics lectures from the 'back door.' Only the second woman to win a Nobel Prize in medicine, Yalow strongly advocated equal opportunity for women. At a Stockholm function in her honor, she said in part, "The world cannot afford the loss of the talents of half of its people."

References

1. Guillemin R. Humoral Hypothalamus Control of the Anterior Pituitary: A Study with Combined Tissue Culture. Endocrinology, vol. 57, pp. 599, 1955.

2. Hadley ME. Endocrinology. 2d ed. Englewood Cliffs, N.J.: Prentice-Hall, 1988.

3. Schally AV (with Saffran M, Benfey BG). Stimulation of the Release of Corticotrophin from the Adenohypophysis by a Neurohypophysial Factor. Canadian Journal of Biochemistry and Physiology, vol. 57, pp. 439, 1955.

4. Schally AV. Purification of Thyrotropic Hormone Releasing Factor from Bovine Hypothalamus. Endocrinology, vol. 78, p. 726, 1966.

5. Stone E. A Mme. Curie from the Bronx. The New York Times Magazine. P. 29, April 9, 1978.

6. Wade N. The Nobel Duel: Two Scientists' Twenty-one Year Race to Win the World's Most Coveted Research Prize. Garden City, N.Y.: Doubleday, 1981.

7. Yalow RS (with Berson SA). Quantitative Aspects of Reaction Between Insulin and Insulin-Binding Antibody. Journal of Clinical Investigation, vol. 38, p. 1996, 1959.

8. Yalow RS. An Interview and What's Ahead for Women in Medicine. Interview by Genevieve Millet. Parents Magazine 53, pp. 38-39, January 1978.

9. Yalow RS. Immunoassay of Endogenous Plasma Insulin in Man. Journal of Clinical Investigation, vol. 39, p. 1157, 1960.

10. Yalow RS. Assay of Plasma Insulin in Human Subjects by Immunological Methods. Nature, vol. 184, p. 1648, 1959.

11. Yalow RS. Plasma Insulin Concentration in Non-Diabetic and Early Diabetic Subjects. Diabetes, vol. 9, p. 254, 1960.

1978

☆

Physics
PYOTR LEONIDOVICH KAPITSA
for his basic inventions and discoveries in the
area of low-temperature physics, and
ARNO A. PENZIAS and **ROBERT W. WILSON**
for their discovery of cosmic microwave background radiation.

☆

Chemistry
PETER D. MITCHELL
for his contribution to the understanding of biological energy
transfer through the formulation of the chemiosmotic theory.

☆

Physiology or Medicine
WERNER ARBER, DANIEL NATHANS,
and **HAMILTON O. SMITH**
for the discovery of restriction enzymes and their
application to problems of molecular genetics.

☆

Literature
ISAAC BASHEVIS SINGER
for his impassioned narrative art which, with roots in a Polish-Jewish
cultural tradition, brings universal human conditions to life.

☆

Peace
MOHAMED ANWAR AL-SADAT
President of the Arab Republic of Egypt. and
MENACHEM BEGIN
Prime Minister of Israel, for jointly negotiating
peace between Egypt and Israel.

☆

Economics
HERBERT A. SIMON

for his pioneering research into the decision-making process within economic organizations.

1978

WERNER ARBER (BORN 1929), DANIEL NATHANS (1928-1999), AND HAMILTON O. SMITH (BORN 1931)

| Werner Arber | Daniel Nathans | Hamilton O. Smith |

© The Nobel Foundation

Werner Arber (left), **Daniel Nathans** (right), **and Hamilton O Smith** shared the 1978 Nobel Prize in Physiology or Medicine "for the discovery of restriction enzymes and their applications in molecular biology."

Born in Gränichen, Aargau, Switzerland, Werner Arber studied at the Swiss Polytechnic School in Zurich and later at the University of Geneva. He spent many years as post-doctoral fellow in the United States before returning to Geneva in 1960.

For his first journal club presentation, Arber chose Watson and Crick's paper on the structure of DNA, which kindled his interest in genetics research. He focused on the process of gene transfers among bacteria with phage as vector. He and his associate Daisy Dussoix worked on radiation-induced damage to genetic materials using *Escherichia coli*, and in the process, discovered restriction enzymes.

Hamilton Smith confirmed Arber's observations. A New Yorker

graduating from the Johns Hopkins Medical School, Smith worked there with Daniel Nathans and discovered a specific restriction enzyme in the *Hemophilus influenza* bacterium. He showed that the enzyme degraded foreign DNA into large, 1000 base-pair fragments without damaging the host DNA. These fragments contained identical sequences of three base-pairs at each end, proving the site of enzyme action to be at the center of specific sequences of six base-pairs on the viral DNA.

Son of Jewish immigrants from Russia to the United States, Daniel Nathans was born in Wilmington, Delaware. He lived in a modest 'cold and leaky' house during childhood and worked at odd jobs during school years to supplement family income. He studied medicine at Washington University in St. Louis and later headed a one-man "Division of Genetics" at the Johns Hopkins medical school.

Nathans pioneered the application of restriction enzymes. In a 1971 paper he reported that the restriction enzyme discovered by Smith cleaved a simian virus SV40 into 11 fragments. Using restriction enzymes, he then deciphered the complete genetic sequence of SV40 and made brilliant predictions about the general use of restriction enzymes in mapping the gene structure. Thus, these scientists formulated the basic tenets of genetic engineering with which restriction enzymes began to be used to determine the order of genes in chromosome and to manufacture 'designer genes.'

After learning her father had won the big prize, Arber's 8-years-old daughter Sylvia asked him to describe his research in simple terms and then made up her own version of the discovery: "...on the tables in my father's laboratory, there are plates with colonies of bacteria, like a city with many people. Inside each bacterium there is a king called 'DNA,' who is very long and skinny. The king has many servants called 'enzymes,' who are thick and short. One such servant serves as a pair of scissors. If a foreign king invades the bacterium, this servant can cut him up into small fragments but does not harm his own king. My father received the Nobel Prize for the discovery of the servant with scissors."

Smith played the piano but claims that he was 'no way gifted' at it. However, when he was thirteen, he heard a recording Beethoven's

Pathetique Sonata performed by Artur Rubinstein—this was to awaken in Smith a lifelong passion for the dramatic beauty of classical music.

References

1. Arber W. Tranduction of Chromosomal Genes and Episomes in E. coli. Virology, vol. 11, pp. 273-288, 1960; (with Dussoix DJ) Host Specificity of DNA Produced by Escherichia coli: I, Host-Controlled Modification of Bacteriophage Lambda. Journal of Molecular Biology, vol. 5, pp. 18-36, 1962; Controlled Modification of Bacteriophages. Annual Review of Microbiology, vol. 19, p. 365-378, 1965; (with Linn S). Host-Controlled Modification of Bacteriophage. Annual Review of Biochemistry, vol. 38, pp. 467-500, 1969.
2. Darnell J, Lodish H, Baltimore D. Molecular Cell Biology. New York: Scientific American Books, 1986.
3. Nathans D (with Danna K). Specific Cleavage of Simian Virus 40 DNA by Restriction Endonuclease of Hemophilus influenzae. Proceedings of the National Academy of Sciences, vol. 68, pp. 2913-2917, 1971. Studies of Simian Virus 40 DNA: (with Danna K, Sack G) VII, A Cleavage Map of the SV40 Genome. Journal of Molecular Biology, vol. 78, pp. 363-376, 1973. Restriction Endonucleases in the Analysis and Restructuring of DNA Molecules. Annual Review of Biochemistry, vol. 46, pp. 273-293, 1975.
4. Smith HO (with Wilcox K). A Restriction Enzyme from Hemophilus influenzae: I, Purification and General Properties. Journal of Molecular Biology, vol. 51, pp. 379-391, 1970; (Kelly TJ) A Restriction Enzyme from Hemophilus influenzae: II, Base Sequence and Recognition Site. Journal of Molecular Biology, vol. 51, pp. 393-409, 1970.

1979

☆

Physics
SHELDON L. GLASHOW, ABDUS SALAM
and STEVEN WEINBERG

for their contributions to the theory of the unified weak and electromagnetic interaction between elementary particles, including inter alia the prediction of the weak neutral current.

☆

Chemistry
HERBERT C. BROWN and GEORG WITTIG

for their development of the use of boron- and phosphorus-containing compounds, respectively, into important reagents in organic synthesis.

☆

Physiology or Medicine
ALAN M. CORMACK and SIR GODFREY N. HOUNSFIELD

for the development of computer-assisted tomography.

☆

Literature
ODYSSEUS ELYTIS (penname of ODYSSEUS ALEPOUDHELIS)

for his poetry, which, against the background of Greek tradition, depicts with sensuous strength and intellectual clear-sightedness modern man's struggle for freedom and creativeness.

☆

Peace
MOTHER TERESA

India, Leader of the Order of the Missionaries of Charity.

☆

Economics
THEODORE W. SCHULTZ and SIR ARTHUR LEWIS

for their pioneering research into economic development research with particular consideration of the problems of developing countries.

1979

Allan MacLeod Cormack (1924-1998), Sir Godfrey Newbold Hounsfield (1919-2004)

Allan MacLeod Cormack   `Sir Godfrey Newbold Hounsfield

© The Nobel Foundation

Allan MacLeod Cormack (left), and **Sir Godfrey Newbold Hounsfield** were awarded the 1979 Nobel Prize for their independent work leading to the development of computerized axial tomography and advanced radiology techniques.

Allan MacLeod Cormack was born in Johannesburg. After a master's degree from the University of Cape Town in 1945 he did postgraduate work at St. John's College, Cambridge, returned to South Africa in 1950, and in 1966 he emigrated to the United States.

While he worked as part-time medical physicist at the University of Cape Town, Cormack was required to compute the attenuation coefficient of X-rays through soft tissues to assess radiation doses for cancer treatment. During a sabbatical at Harvard University using a desktop calculator, he worked out mathematical solutions for this problem. Although he reconstructed cross sectional images of irregularly shaped objects accurately, in the early 1960s, there was no X-ray equipment to test his equations in practical scenarios.

Sir Godfrey Hounsfield was born on a farm in Nottinghamshire, England. After service in the Royal Air Force between 1939 and 1946, he studied at Faraday House Electric Engineering College in London and obtained a degree in engineering. He worked at the Electrical and Musical Instrument (EMI) Ltd., directing research and development on electronic products for commercial use.

Initially, Hounsfield contributed to the development of the first solid-state computer manufactured in Great Britain. He then proposed to the EMI, to develop computer-based X-ray scanners. He realized that computer's abilities to recognize printed characters could be adapted to computing the X-ray attenuation coefficients through soft tissues. His work helped develop the famous "EMI-scanner." A prototype clinical scanner was first installed in 1971 in Atkinson Morley's Hospital, Wimbledon, England. Hundreds of CT scanners became available worldwide within a decade, making a profound impact on medical practice at large and on neurology.

Since the 1901 Nobel Prize in Physics awarded to Roentgen for the discovery of X-rays, the awards given to Sir Godfrey and Cormack were the first for an invention in diagnostic radiology honored in the category of Physiology or Medicine.

References

1. Hounsfield GN (with Brown PH). Magnetic Films for Information Storage. British Patent Specification 1083673, September 20, 1967.
2. Hounsfield GN. A Method of and Apparatus for Examination of a Body by Radiation Such as X or Gamma Radiation. British Patent Specification 1283915, August 1972.

1980

☆

Physics
JAMES W. CRONIN and VAL L. FITCH
*for the discovery of violations of fundamental symmetry
principles in the decay of neutral K-mesons.*

☆

Chemistry
PAUL BERG
*for his fundamental studies of the biochemistry of nucleic
acids, with particular regard to recombinant-DNA and*
WALTER GILBERT and FREDERICK SANGER
*for their contributions concerning the determination
of base sequences in nucleic acids.*

☆

Physiology or Medicine
BARUJ BENACERRAF, JEAN DAUSSET and GEORGE D. SNELL
*for their discoveries concerning genetically determined structures
on the cell surface that regulate immunological reactions.*

☆

Literature
CZESLAW MILOSZ
*who with uncompromising clear-sightedness voices man's
exposed condition in a world of severe conflicts.*

☆

Peace
ADOLFO PEREZ ESQUIVEL
Argentina, architect, sculptor and human rights leader.

☆

Economics
LAWRENCE R. KLEIN
*for the creation of econometric models and the application to the
analysis of economic fluctuations and economic policies.*

1980

| George Davis Snell | Jean Baptiste Dausset | Baruj Benacerraf |

© The Nobel Foundation

George Davis Snell (left), **Jean Baptiste Dausset** (right), and **Baruj Benacerraf** were awarded the 1980 Nobel Prize in Physiology or Medicine for their independent contributions to the field of immunology.

George Snell was born in Bradford, Massachusetts and studied at Dartmouth College and Harvard University. In 1935, he joined Jackson Memorial Laboratory at Bar Harbor, Maine, as researcher. Fascinated by the phenomenon of recognition of "self" by the body, he conducted transplantation studies in mice that spanned over two decades. By repeated breeding he developed "congenic mice"—mice that were genetically identical except for a single locus. He found ten loci controlling graft resistance, one of which proved to be of great importance in graft acceptance. Named the histocompatibility II (H-2) locus, it was latter called the major histocompatibility complex (MHC).

Son of a famous French physician, Jean Dausset was born in

273

Toulouse. During World War II, he was a part of the Free French campaign in North Africa and Normandy. In 1945, he earned his M.D., from the University of Paris and joined the National Blood Transfusion Center in Paris. In studying patients who had received multiple transfusions, Dausset discovered unusual antibodies, which he showed to be caused by human leukocyte antigens (HLA). HLA, one of the MHC antigens, was the human equivalent of Snell's mice H-2. Dausset suggested that HLA were localized on a single chromosome, later shown to be chromosome 6.

Born in Caracas, Venezuela, Baruj Benacerraf grew up in Paris. The impending war in 1938 forced the family to move back to Venezuela. Young Benacerraf then pursued his studies in the USA. Because he was a Jew and a foreigner, he was rejected by 25 medical schools before being accepted by the Medical College of Virginia, from where he graduated in 1945. From 1956-68 he spent a most fruitful period of his research career at the New York University School of Medicine.

Benacerraf became interested in immunology because of asthma he suffered since childhood. As he attempted to produce pure antibodies using DNP-poly-L-lysin antigens, he and his colleagues discovered that specific antibody production was controlled by specific set of genes—the immune response (Ir) genes. These "super genes," located within the MHC were shown to be of great importance in controlling communication among the immunocytes responsible for body's immune response.

The discoveries of these scientists became the foundation for transplantation immunology and the modern era of organ transplantation.

References

1. Benacerraf B 9 (with Gell PGH). Studies on Hypersensitivity: II, Delayed Hypersensitivity to Denatured Proteins in Guinea Pigs. Immunology, vol. 2, p. 64, 1959; Studies on Artificial Antigens: (Levine BB, Ojeda A) III, The Genetic Control of The Immune

Response to Hapten-poly-L-lysine Conjugates in Guinea Pigs. Journal of Experimental Medicine, vol. 118, pp. 953-957, 1963.

2. Benacerraf B. The Training Experience. Journal of Immunology, vol. 113, pp. 431-437, 1974; Son of an Angel, 1991.

3. Bibel DJ. Milestones in Immunology: A Historical Exploration. Madison, Wis.: Science Tech, 1988

4. Cohen IR. The Self, the World, and Autoimmunity. Scientific American 258, pg. 52-60, February 1988.

5. Cunningham BA. The Structure and Function of Histocompatibility Antigens. Scientific American 237, pg. 96-107, October 1977.

6. Dausset J. Immunohématologie biologique et clinique, 1956; (with Rapaport FT) Human Transplantation, 1968; (with Colombian J) Histocompatibility Testing: 1972, 1972; (with Svejgaard A) HLA and Disease; (with Snell G, Nathenson S) Histocompatibility, 1976; (with Rapaport FT) A Modern Illustration of Experimental Medicine in Action, 1980; Tissue Typing, 1966, Tissue Typing Today, 1971.

7. Snell GD (with the staff of the Jackson Laboratory). Biology of the Laboratory Mouse, 1941; The Induction by X-Rays of Heredity Changes in Mice. Genetics, vol. 20, pp. 545-567, 1935.

1981

☆

Physics
NICOLAAS BLOEMBERGEN and ARTHUR L. SCHAWLOW
for their contribution to the development of laser spectroscopy and
KAI M. SIEGBAHN
for his contribution to the development of high- resolution electron spectroscopy.

☆

Chemistry
KENICHI FUKUI and ROALD HOFFMANN
for their theories, developed independently, concerning the course of chemical reactions.

☆

Physiology or Medicine
ROGER W. SPERRY
for his discoveries concerning the functional specialization of the cerebral hemispheres, and
DAVID H. HUBEL and TORSTEN N. WIESEL
for their discoveries concerning information processing in the visual system.

☆

Literature
ELIAS CANETTI
for writings marked by a broad outlook, a wealth of ideas and artistic power.

☆

Peace
OFFICE OF THE UNITED NATIONS HIGH COMMISSIONER FOR REFUGEES
Geneva, Switzerland.

☆

Economics
JAMES TOBIN
for his analysis of financial markets and their relations to expenditure decisions, employment, production and prices.

1981

ROGER WOLCOTT SPERRY (1913-1994);
DAVID HUNTER HUBEL (B 1926); AND
TORSTEN N. WIESEL (BORN 1924).

Roger Wolcott Sperry   David Hunter Hubel   Torsten Nils Wiesel

© The Nobel Foundation

One-half of the 1981 Nobel Prize in Physiology or Medicine was awarded to **Roger Wolcott Sperry** (left), for his discoveries concerning "the functional specialization of cerebral hemispheres," and the other half jointly to **David Hubel** (right), and **Torsten Wiesel** "for their discoveries concerning the visual system."

Born in Hartford, Connecticut, Sperry studied zoology at the University of Chicago. Since 1954, he worked at the California Institute of Technology. Exploring the electrical field theory pertaining to sensory integration and cognition, Sperry experimented on cats, and later monkeys. His early work showed that after severing the corpus callosum (commissurotomy), these animals developed major cognitive disturbances—contradicting the prevailing view that corpus callosal connections were of little importance.

Extending this work in patients who had undergone split-brain surgery for intractable epilepsy, Sperry and colleagues showed that after commissurotomy, each half of the brain worked as if being 'outside the conscious realm of the other.' Besides exploring the role of cortical connections, in an illustrious career spanning over four decades, Sperry formulated many fundamental concepts in neurobiology and psychobiology.

David Hubel was born in Windsor, Ontario, Canada and studied at McGill University. While he was a resident at Johns Hopkins Hospital in the USA, he was drafted by the US Army and posted at the Walter Reed Army Institute of Research. In 1958, he rejoined Hopkins for a fellowship in Ophthalmology. There, Hubel began his decade-long collaborative work with Torsten Wiesel, a postdoctoral fellow from Sweden who had graduated in medicine from the Karolinska Institute. According to George Berry of Harvard University, "They were like twins."

By suturing individual eyelids for varying periods in experimental cats and monkeys, Hubel and Wiesel tested the role of visual experience in visual cognition. They used microelectrode techniques to record the neuronal signals in the midbrain and the visual cortex and mapped cortical visual fields. Among other things, they discovered the process of visual image encoding in the cortex and showed that there were critical periods of sensory development in the brain.

Like the deciphering of the Egyptian hieroglyphics, Hubel and Wiesel's discoveries have been hailed as one of the most important attempts to describe the "secrets of the brain."

References

1. Begbie GH. Seeing and the Eye: An Introduction to Vision. Garden City. N.Y.: Natural History Press, 1969.
2. Hubel DH (with Wiesel T). Cortical Unit Responses to Visual Stimuli in Non-anesthetized Cats. American Journal of Ophthalmology, vol. 46, pp. 110-122, 1958; Receptive Fields of Single Neurons in the Cat's Striate Cortex. Journal of Physiology,

vol. 148, pp. 574-591, 1959; Shape and Arrangement of Columns in Cat's Striate Cortex. Journal of Physiology, vol. 165, pp. 559-568, 1963.

3. Hubel DH. Mystery of Matter and Mind. New York: Torstar Books, 1984.

4. Sperry RW. Cerebral Organization and Behaviour. Science, vol. 133, pp. 1749-1757, 1961; (with Gazzaniga MS) Language Following Disconnection of the Hemispheres. Brain Mechanisms Underlying Speech and Language, edited by C.H. Millikan and F.L. Darley; Language After the Section of Cerebral Commissures. Brain, vol. 90, pp. 131-148, 1967; (with Saul R) Absence of Commissurotomy Symptoms with Agenesis of the Corpus Callosum. Neurology, vol. 18, p. 307, 1968.

5. Wiesel TN (with Hubel D). Ferrier Lecture: Functional Architecture of Macaque Monkey Visual Cortex. Proceedings of the Royal Society of London, series B, vol. 198, pp. 1-59, 1977.

6. Wiesel TN. Postnatal Development of the Visual Cortex and the Influence of Environment. Nature, vol. 299, pp. 583-591, 1982.

1982

☆

Physics
KENNETH G. WILSON
for his theory for critical phenomena in connection with phase transitions.

☆

Chemistry
SIR AARON KLUG
for his development of crystallographic electron microscopy and his structural elucidation of biologically important nuclei acid-protein complexes.

☆

Physiology or Medicine
SUNE K. BERGSTRÖM, BENGT I. SAMUELSSON and SIR JOHN R. VANE
for their discoveries concerning Prostaglandins and related biologically active substances.

☆

Literature
GABRIEL GARCÍA MÁRQUEZ
for his novels and short stories, in which the fantastic and the realistic are combined in a richly composed world of imagination, reflecting a continent's life and conflicts

☆

Peace
ALVA MYRDAL
former Cabinet Minister, diplomat, delegate to United Nations General Assembly on Disarmament, writer; and
ALFONSO GARCÍA ROBLES
Diplomat, delegate to the United Nations General Assembly on Disarmament, former Secretary for Foreign Affairs.

☆

Economics
GEORGE J. STIGLER

for his seminal studies of industrial structures, functioning of markets and causes and effects of public regulation.

1982

KARL SUNE DETLOF BERGSTRÖM
(1916-2014), BENGT INGEMAR
SAMUELSSON (2034-2024) AND SIR
JOHN ROBERT VANE (1927-2004)

Karl Sune Detlof Bengt Ingemar Sir John Robert Vane
Bergström Samuelsson

© The Nobel Foundation

Sune Bergström (left), **Bengt Samuelsson** (right), and **John Vane** shared the 1982 Nobel Prize in Physiology or Medicine "for their research concerning prostaglandins."

Bergström was born in Stockholm and studied medicine from Royal Karolinska Institute. After a brief research experience in England and the USA, he rejoined Karolinska Institute as faculty and pursued lipid research under Ulf von Euler (the 1970 Nobel laureate), who had discovered "prostaglandin" from the secretions of sheep prostate glands.

Bergström realized that the study of prostaglandin biochemistry posed formidable challenges because of their small quantities. By the late 1950s, Ragnar Ryhage, a chemist, and Bengt Sammuelsson, Bergström's graduate student had developed a new instrument that combined gas chromatography and mass spectrometer, greatly aiding prostaglandin measurements. By 1962, Bergström, Sammuelsson and colleagues discovered that there were at least six compounds that made up the prostaglandin-complex, and arachidonic acid was their common precursor.

John Vane was born in Tardebigg, Worchestershire, UK, and studied in Oxford University earning his Ph.D. in 1953. Vane credits Harold Burn for inspiring in him an interest in pharmacology. For brief periods, Vane worked at Yale University and at the University of London before joining in 1973, the Wellcome Foundation as Director of Research. Vane's first major contribution was in the field of bioassay. He developed such advanced laboratory methods as superfusion bioassay, with which up to six tissues could be studied simultaneously, and blood-bathed organ technique with which the biological properties of tiniest amounts of chemical substances could be assessed.

After a series of studies, Vane and colleagues discovered that during anaphylaxis in the guinea pigs three substances are released— prostaglandin E_2, $F_{2\alpha}$, and a hitherto unknown chemical, prostacyclin. Vane also discovered that aspirin blocked prostacyclin actions. On a weekend in 1971, he formulated a brilliant hypothesis that the actions of aspirin were mediated by its inhibitory effect on prostaglandin. Within a week, he had proved it.

The ubiquitous nature of prostaglandins and the discovery of their physiological properties led to the deciphering of the mysteries of such universally common symptoms as fever, pain, and inflammation. Prostaglandin research also led to the development of treatments for diverse conditions. Prostaglandins and their inhibitors are now used routinely in stopping premature uterine contractions during pregnancy, closure or maintenance of the patency of the ductus arteriosus in newborn infants, prevention of coronary heart disease in adults, and in many hypertensive disorders associated with pregnancy.

References

1. Bergström SKD (with Sjövall J). The Isolation of Prostaglandin. Acta Chemica Scandinavia, vol. 11, p. 1086, 1957; The Isolation of Prostaglandin F from Sheep Prostate Glands. Acta Chemica Scandinavica, vol. 14, pp. 1693-1700, 1960.

2. Bergström SKD. The Enzymatic Conversion of Essential Fatty Acids into Prostaglandins. Journal of Biological Chemistry, vol. 239, p. 4006-4008, 1964.

3. Vane JR (with Bakhle YS). Metabolic Functions of the Lung. 1977; Prostacyclin in Health and Disease, 1982.

4. Vane JR (with Bergström S). Anti-Inflammatory Drugs, 1979; Prostacyclin, 1979.

5. Vane JR. Inhibition of Prostaglandin Synthesis as a Mechanism of Action for Aspirin-Like Drugs. Nature: New Biology, June 23, 1971.

1983

☆

Physics
SUBRAMANYAN CHANDRASEKHAR
*for his theoretical studies of the physical processes of importance
to the structure and evolution of the stars,* and
WILLIAM A. FOWLER
*for his theoretical and experimental studies of the nuclear reactions of
importance in the formation of the chemical elements in the universe.*

☆

Chemistry
HENRY TAUBE
*for his work on the mechanisms of electron transfer
reactions, especially in metal complexes.*

☆

Physiology or Medicine
BARBARA MC CLINTOCK
for her discovery of mobile genetic elements.

☆

Literature
SIR WILLIAM GOLDING
*for his novels which, with the perspicuity of realistic narrative
art and the diversity and universality of myth, illuminate
the human condition in the world of today.*

☆

Peace
LECH WALESA
Poland. Founder of Solidarity, campaigner for human rights.

☆

Economics
GERARD DEBREU
*for having incorporated new analytical methods into economic theory and
for his rigorous reformulation of the theory of general equilibrium.*

1983

BARBARA MCCLINTOCK (1902-1992)

© The Nobel Foundation

Barbara McClintock won the Nobel Prize for Physiology or Medicine "for her discovery of mobile genetic elements."

Daughter of a physician, she was born in Hartford Connecticut. She grew up in rural New England and enjoyed nature, and "thinking about things." Despite her mother's initial objections about the idea of her going to college, Barbara joined Cornell University and majored in botany earning a Ph.D., in 1927. She worked at the University of Missouri from 1936 to 1941 and took a position at Cold Spring Harbor Laboratory, where she remained for most of her career.

Even from her Cornell days, McClintock was fascinated by cytogenetics. Lewis Stadler at the University of Missouri suggested her to study the mutations he had induced in maize by X-rays. In those, McClintock discovered loss of chromosomes, rearrangements, fusion of broken ends, and ring chromosomes. Pursuing this line of research at Cold Spring Harbor, she studied seedling mutations in about 450 plants, whose parents each had a newly ruptured chromosome arm. She concluded that gene expression was regulated by events during mitosis, and that "transposable elements," either gained or lost, played the most important role in the process.

By 1947, McClintock began exploring the mechanisms underlying genetic instability that was responsible for causing altered patterns of pigmentation in maize kernels. Her work established that variegation in the color of the kernels paralleled the transposition of structural elements within or between chromosomes. When she presented this theory of transposition of genes at a symposium, it was not understood or accepted by her colleagues. "They called me crazy, absolutely mad at times," later she would recall dryly.

McClintock had made most of her groundbreaking discoveries at a time when neither the nature of the genetic code nor the structure of the DNA molecule had been discovered. Years later it was shown that McClintock's "jumping gene" theory of recombination of the DNA segments was central to explaining such diverse functions as the lymphoid cell immune response and transformation of normal cells into cancer cells.

McClintock faced sexual bias all her life. According to her, the University of Missouri was particularly "awful, awful, awful." The chairman of the botany department told her that she would never secure a tenure position, and if she got married, she would be fired. At a press conference after the Nobel Prize she said, "You don't need the public recognition; you just need the respect of your colleagues."

References

1. Keller EF. A Feeling for the Organism: The Life and Work of Barbara McClintock. New York: W.H. Freeman, 1983.
2. McClintock B. A Cytological and Genetic Study of Triploid Maize. Genetics, vol. 14, pp. 180-222, 1929; (with Creighton H) A Correlation of Cytological and Genetic Crossing Over in Zea Mays. Proceedings of the National Academy of Sciences, vol. 17, pp. 492-497, 1931; (with Creighton H) Proceedings of the National Academy of Sciences, vol. 17, pp. 492-497, 1931; A Correlation

of Ring-Shaped Chromosomes with Variegation in Zea Mays. Proceedings of the National Academy of Sciences, vol. 18, pp. 677-681, 1932; The Stability of Broken Ends of Chromosomes in Zea Mays. Genetics, vol. 26, pp. 234-282, 1941; The Association of Mutants with Homozygous Deficiencies in Zea Mays. Genetics, vol. 26, pp. 542-571, 1941; Cytogenetic Studies of Maize and Neurospora. Carnegie Institute of Washington Year Book, vol. 44, p. 108, 1945.

1984

☆

Physics
CARLO RUBBIA and SIMON VAN DER MEER
for their decisive contributions to the large project, which led to the discovery of the field particles W and Z, communicators of weak interaction.

☆

Chemistry
ROBERT BRUCE MERRIFIELD
for his development of methodology for chemical synthesis on a solid matrix.

☆

Physiology or Medicine
NIELS K. JERNE, GEORGES J.F. KÖHLER and CÉSAR MILSTEIN
for theories concerning the specificity in development and control of the immune system and the discovery of the principle for production of monoclonal antibodies.

☆

Literature
JAROSLAV SEIFERT
for his poetry which endowed with freshness, sensuality and rich inventiveness provides a liberating image of the indomitable spirit and versatility of man.

☆

Peace
DESMOND MPILO TUTU
South Africa, Bishop of Johannesburg, former Secretary General South African Council of Churches (S.A.C.C.). for his work against apartheid

☆

Economics
SIR RICHARD STONE
for having made fundamental contributions to the development of systems of national accounts and hence greatly improved the basis for empirical economic analysis.

1984

NIELS KAJ JERNE (1911-1994), GEORGES JEAN FRANZ KÖHLER (1946-1995) AND CÉSAR MILSTEIN (1926-2002)

Niels Kaj Jerne   **Georges Jean Franz Köhler**   **César Milstein**

© The Nobel Foundation

Niels Kai Jerne (left), **Georges Jean Franz Köhler** (right), and **César Milstein** shared the 1984 Nobel Prize in Physiology or Medicine for their work in immunology that led to the discovery of the principles of monoclonal antibody production.

Jerne was born in London to Danish parents and studied in The Netherlands and in Denmark. While a medical student in the University of Copenhagen, he worked as researcher at the Danish State Serum Institute. Later, Jerne held posts at the WHO in Geneva and in other academic and research institutions in Pittsburgh, Frankfurt, Basel, and Paris.

Considered "the most intelligent immunologist" of his time, Jerne was a great theoretician. In 1955 he proposed the "natural-selection theory of antibody formation," according to which everyone produces many 'natural' antibodies during fetal life, independent of exogenous stimulation. In 1971 Jerne proposed the "somatic generation of immune recognition" theory that explained the process of development of the immune system from stem cells to mature lymphocytes. He proposed that immunological specificity was regulated through genes.

Based on the observation that antibodies elicited 'anti-antibodies', Jerne proposed his most important theory—the "network theory." According to this, the anti-antibodies can produce 'anti-anti-antibodies,' leading to an endless cascade of 'network' of antibodies. Jerne's fundamental contributions in theoretical immunology greatly influenced research in applied immunology.

Born in Bahíya Blanka, Argentina, Milstein studied biochemistry there and later did research under Frederick Sanger at Cambridge. Milstein returned to Buenos Aires in 1962, but political upheaval forced him to leave Argentina in 1963 and join the MRC Laboratory of Molecular Biology in Cambridge. In 1974, Georges Köhler, a Munich-born scientist, joined Milstein as postdoctoral fellow.

Köhler and Milstein tried to produce a clone of cells that could generate a single antibody. In 1975, the pair developed the "hydridoma technique," which involved fusion of an antibody producing cell line with a mouse myeloma cell line for 'immortalizing' the antibody production.

Representing the MRC, Milstein contacted the British patent office to see if the latter wished to patent the hybridoma technique—the office did not respond. Köhler and Milstein explained the technique to the world in a short article in *Nature* (1975; 256:495-7). The paper was sensational. Thus, the British government lost a chance of profiting from the hybridoma technique, with which unlimited production of single lines of antibodies with predetermined specificity became possible.

In 1976, Köhler left the MRC. A man of high ethical standards, Köhler refused to become an 'antibody industrialist.' Instead, he pursued fundamental research in immunology in Switzerland and in Germany.

References

1. Cherfas J. Monoclonals: A New Industry Is Born. New Scientist 104, October 18, 1984.

2. Goding JW. Monoclonal Antibodies: Principles and Practice. New York: Academic Press, 1986.

3. Jerne NK. The Natural Selection Theory of Antibody Formation. Proceedings of the National Academy of Sciences, USA, vol. 41, pp. 849-856, 1955; Immunological Speculations. Annual Review of Microbiology, vol. 14, pp. 341-358, 1960.

4. Jerne NK. The Natural Selection Theory of Antibody Formation: Ten Years Later. Phage and the Origins of Molecular Biology, pp. 301-312, 1966.

5. Köhler GJF (with Milstein). Continuous Cultures of Fused Cells Secreting Antibody of Predefined Specificity. Nature, vol. 256, pp. 495-497, August 7, 1975. Immunoglobulin Chain Loss in Hybridoma Lines. Proceedings of the National Academy of Sciences, USA, vol. 77, pp. 2197-2199, 1980; (with Shulman MJ, Heusser C, Filkin C). Mutations Affecting the Structure and Function of Immunoglobulin M. Molecular and Cellular Biology, vol. 2, pp. 1033-1043, 1982.

6. Milstein C (with Brenner S). Origins of Antibody Variation. Nature, vol. 211, pp. 242-243, 1966; Linked Groups of Residues in Immunoglobulin K Chains. Nature, vol. 216, pp. 330-332, 1967; (with Adetugbo K) Molecular Analysis of Spontaneous Somatic Mutants. Nature, vol. 265, pp. 299-304, 1977.

1985

☆

Physics
KLAUS VON KLITZING
for the discovery of the quantized Hall effect.

☆

Chemistry
HERBERT A. HAUPTMAN and JEROME KARLE
for their outstanding achievements in the development of direct methods for the determination of crystal structures.

☆

Physiology or Medicine
MICHAEL S. BROWN and JOSEPH L. GOLDSTEIN
for their discoveries concerning the regulation of cholesterol metabolism.

☆

Literature
CLAUDE SIMON
who in his novel combines the poet's and the painter's creativeness with a deepened awareness of time in the depiction of the human condition.

☆

Peace
INTERNATIONAL PHYSICIANS FOR THE PREVENTION OF NUCLEAR WAR
Boston, MA, U.S.A.

☆

Economics
FRANCO MODIGLIANI
for his pioneering analyses of saving and of financial markets.

1985

JOSEPH LEONARD GOLDSTEIN (B 1940),
AND MICHAEL STUART BROWN (B 1941)

Joseph Leonard Goldstein   **Michael Stuart Brown**

© The Nobel Foundation

Joseph Leonard Goldstein (left), and **Michael Stuart Brown** shared the 1985 Nobel Prize for Physiology or Medicine for their work related to cholesterol metabolism.

Born in Sumter, South Carolina, Goldstein obtained his MD degree in 1966 from the University of Texas, Southwestern Medical School in Dallas. During an internship at Boston's Massachusetts General Hospital, he met and befriended a fellow-intern Michael Brown—a New Yorker, who had earned his MD degree in 1966 from the University of Pennsylvania in Philadelphia. After their respective residency training, both physicians joined the National Institutes of Health in Bethesda, Maryland, and worked as clinical associates—Goldstein in Biochemical Genetics and Brown in Digestive and Hereditary Diseases.

In 1971, Brown joined the division of gastroenterology in

Southwestern Medical school in Dallas. The following year, Goldstein also moved to his alma matter and became the head of the new Division of Medical Genetics. These scientists began collaborative research almost immediately, studying patients with familial hypercholesterolemia (FH). They combined their laboratories in 1974.

In 1973, Brown, Goldstein and coworkers discovered that in homozygous FH patients the uptake of low-density lipoproteins (LDL) was markedly deficient, while it was moderately deficient in heterozygous patients. They hypothesized that the defects in LDL uptake were due to the deficiency of receptors on cell surface, secondary to single-gene mutation. In 1976, the team reported localizing the LDL receptors; they were the "coated pits" on cell surfaces, which were deficient or nonexistent in FH patients.

Brown and Goldstein also proposed that LDL uptake and internalization occurred through the process of receptor-mediated endocytosis, thereby modulating cholesterol synthesis. In the late 1980s, the human LDL receptor gene was sequenced and shown to be a mosaic of many exons (domains). Similarly, several mutations were found for the LDL receptor gene.

The discovery of the LDL-receptors by Brown and Goldstein was a milestone in cholesterol research. Besides helping patients with FH, an understanding of how normal cells handle lipids, cholesterol and related molecules was pivotal in exploring the intricate relationship between lipids and atherosclerosis in otherwise normal individuals.

References

1. Brown MS, Goldstein JL (with Schneider and Beisiegal U). Purification of the Low-Density Lipoprotein Receptor, an Acidic Glycoprotein of 164,000 Molecular Weight. Journal of Biological Chemistry, vol. 257, pp. 2664-2673, 1982.

2. Brown MS, Goldstein JL. Expression of the Familial Hypercholesterolemia Gene in Heterozygotes: Mechanism for a Dominant Disorder in Man. Science, vol. 185, pp. 61-63,

1974; The LDL Receptor Locus and the Genetics of Familial Hypercholesterolemia. Annual Review of Genetics, vol. 13, pp. 259-289, 1979.

3. Brown MS, Goldstein JL. The Low-Density Lipoprotein Pathway and Its Relation to Atherosclerosis. Annual Review of Biochemistry, vol. 46, pp. 897-930, 1977; Atherosclerosis: The Low-Density Lipoprotein Receptor Hypothesis. Metabolism, vol. 26, pp. 1257-1275, 1977; Familial Hypercholesterolemia: Model for Genetic Receptor Disease. The Harvey Lectures Series, vol. 73, pp. 163-201, 1979.

1986

☆

Physics
ERNST RUSKA
*for his fundamental work in electron optics, and for the
design of the first electron microscope,* and
GERD BINNIG and HEINRICH ROHRER
for their design of the scanning tunneling microscope.

☆

Chemistry
DUDLEY R. HERSCHBACH, YUAN T. LEE and JOHN C. POLANYI
*for their contributions concerning the dynamics
of chemical elementary processes.*

☆

Physiology or Medicine
STANLEY COHEN and RITA LEVI-MONTALCINI
for their discoveries of growth factors.

☆

Literature
WOLE SOYINKA
*who in a wide cultural perspective and with poetic
overtones fashions the drama of existence.*

☆

Peace
ELIE WIESEL
Chairman of 'The President's Commission on
the Holocaust.' Author, humanitarian.

☆

Economics
JAMES M. BUCHANAN, JR.
*for his development of the contractual and constitutional bases
for the theory of economic and political decision-making.*

1986

Rita Levi-Montalcini   **Stanley Cohen**
© The Nobel Foundation

Rita Levi-Montalcini and **Stanley Cohen** received the 1986 Nobel Prize in Physiology or Medicine 1986 for their discoveries of growth factors.

Stanley Cohen was born in Brooklyn, New York, to Jewish emigrants from Russia. After obtaining a Ph.D., in biochemistry from the University of Michigan in 1945, he worked for a while in Colorado before joining Washington University at St. Louis as postdoctoral fellow in the department of radiology.

Rita Levi-Montalcini and her twin sister Paola were born in Turin, Italy to Jewish parents. In 1936, the year Rita graduated from the medical school in Turin, Mussolini forbade Jews from practicing medicine, forcing her to take up a research career. She began studying neuroembryology, initially in Belgium, later in Turin. Setting up a laboratory in her kitchen using household utensils and a microscope, she studied growth patterns in chick embryos. During World War II, she lived underground in Florence for two years. In 1947, she accepted an invitation for a research position from Viktor Hamburger of Washington University in St. Louis; she spent the rest her career in the USA.

By then, Elmer Bueker had shown that mouse sarcoma tumors, when grafted onto chick embryos, led to the formation of sensory nerve fibers in the host tissue. Expanding on this work, Hamburger and Levi-Montalcini discovered that a chemical factor—the "nerve growth factor" (NGF) — released from the tumor was responsible for that phenomenon.

In 1954, Cohen joined Levi-Montalcini and Hamburger to pursue research on NGF. To destroy the nucleic acid in the preparation of NGF, Cohen used snake venom; to his surprise, he found that the venom also contained potent NGF. Since venom in the snake was like saliva in mammals, he tried and succeeded in extracting NGF from mouse salivary glands. Later, the team purified NGF and produced its antiserum, aiding in studies on the biological functions of NGF.

In 1959, Cohen moved to Vanderbilt University in Nashville, Tennessee, and continued work on growth factors. He had noted that salivary gland extracts, but not pure NGF, enhanced skin growth in newborn mice. By 1962, he had discovered an additional chemical responsible for skin maturation—he named it "epidermal growth factor" (EGF).

It was soon learned that growth factors were needed not only for growth and differentiation of the embryo and the fetus, but also for receptor-mediated hormonal response. Later studies also revealed their structural similarities to oncogenes, suggesting a role for them in tumor pathogenesis. Although their full biological and evolutionary significance remains to be characterized the discovery of NGF and EGF heralded a new era in the study of normal and abnormal growth and differentiation of organs and tissues.

References

1. Bradshaw RA. Nerve Growth Factor. Annual Review of Biochemistry 47, pp. 191-216, 1978.
2. Cohen S (with Taylor J). Epidermal Growth Factor: Chemical and Biological Characterization. Recent Progress in Hormone Research, vol. 30, pp. 533-550, 1974; (with Savage R) Recent Studies on the Chemistry and Biology of Epidermal Growth Factor. Recent Progress in Hormone Research, vol. 30, pp. 551-574, 1974; (with Carpenter G) Epidermal Growth Factor. Annual Review of Biochemistry, vol. 48, 193-216, 1979.
3. Montalcini RL (with Hamburger V). Proliferation Differentiation, and Degeneration in the Spinal Ganglia of the Chick Embryo Under Normal and Experimental Conditions. Journal of Experimental Zoology, vol. 111, pp. 457-502, 1949; Selective Growth Stimulating Effects of Mouse Sarcoma on the Sensory and Sympathetic Nervous System of the Chick Embryo. Journal of Experimental Zoology, vol. 115, pp. 321-362; (with Cohen S) A Nerve Growth Stimulating Factor Isolated from Snake Venom. Proceedings of the National Academy of Sciences, USA, vol. 42, pp. 571-574, 1956.

1987

☆

Physics

J. GEORG BEDNORZ and K. ALEXANDER MÜLLER

*for their important breakthrough in the discovery of
superconductivity in ceramic materials.*

☆

Chemistry

**DONALD J. CRAM, JEAN-MARIE LEHN
and CHARLES J. PEDERSEN**

*for their development and use of molecules with structure-
specific interactions of high selectivity.*

☆

Physiology or Medicine

SUSUMU TONEGAWA

for his discovery of the genetic principle for generation of antibody diversity.

☆

Literature

JOSEPH BRODSKY

*for an all-embracing authorship, imbued with
clarity of thought and poetic intensity.*

☆

Peace

OSCAR ARIAS SANCHEZ

Costa Rica, President of Costa Rica, initiator of
peace negotiations in Central America.

☆

Economics

ROBERT M. SOLOW

for his contributions to the theory of economic growth.

1987

© The Nobel Foundation

Susumu Tonegawa received the 1987 Nobel Prize in Physiology or Medicine "for his discovery of the genetic principle for generation of antibody diversity."

Born in Nagoya, Japan, Tonegawa studied chemistry at Kyoto University. He went on a Fulbright travel grant to the USA for studies in molecular biology and in 1968, earned his Ph.D., from the University of California, San Diego. Since the US visa required him to spend two years abroad after it expired in 1970, he went to Switzerland and worked at the Institute of Immunology in Basel. It was in Basel that he did much of his Nobel Prize-winning work. In 1981, Tonegawa moved back to the USA, accepting a professorship at the Center for Cancer Research of the Massachusetts Institute of Technology.

In the 1960s, it was unclear how the body recognized a vast number of antigens to produce antibodies. One theory held that the germ line was programmed to generate all future antibodies—but this required an enormous number of genes for antibody coding. The other hypothesis, the "two gene-one polypeptide" theory, proposed that only a limited number of genes in the germ line was sufficient to determine antibody diversity, probably by recombination.

Using the new techniques of restriction enzymes, recombinant DNA and Southern blot, Tonegawa and Nobumichi Hozumi tested the latter theory. They used cloned DNA from myeloma cells and DNA from mouse embryo, and found that in the embryonic DNA, the constant and variable segments of the immunoglobulin gene were separate. Tonegawa with Philip Leder discovered a 'missing' DNA segment, the 'J' segment (for joining) of the gene, responsible for joining different segments. One of many J segments and one of many variable segments can generate innumerable immunoglobulin light chains by random recombination and mutation, accounting for antibody diversity.

Later, with Leroy Hood and other colleagues, Tonegawa showed that a third DNA segment—'D' for diversity—was also involved in antibody diversity. Further work by these and other investigators clarified the development and structure of T cell receptors, and the role of DNA rearrangement in lymphomas and leukemia.

Comparing the reshuffling of DNA segments (to form millions of antibodies) to car manufacturing, Tonegawa said... "If [General Motors] custom-made each car, it would not be economical. So, they make different parts, then they reassemble them in different ways [to make numerous models]."

The landmark discoveries of Tonegawa and colleagues provided a deeper insight into the workings of the body's defense system, with implications for treating autoimmune diseases, improving immunological therapy and minimizing graft rejection.

References

1. Tonegawa S (with Nobumichi H). Evidence for Somatic Rearrangement of Immunoglobulin Genes Coding for Variable

and Constant Regions. Proceedings of the National Academy of Science, USA, vol. 73, pp. 3628-3632, 1976.

2. Tonegawa S (with Kurosawa Y). Organization, Structure, and Assembly of Immunoglobulin Heavy Chain Diversity DNA Segments. Journal of Experimental Medicine, vol. 155, pp. 201-218, 1982.

3. Tonegawa S (with Heilig JS). T-Cell Gamma Gene is Allelically but Not Isotopically Excluded and Is Not Required in Known Functional T-Cell Subsets. Proceedings of the National Academy of Sciences, USA, vol. 84, pp. 8070-8074, 1987.

1988

☆

Physics
LEON M. LEDERMAN, MELVIN SCHWARTZ and JACK STEINBERGER
for the neutrino beam method and the demonstration of the doublet structure of the leptons through the discovery of the muon neutrino.

☆

Chemistry
JOHANN DEISENHOFER, ROBERT HUBER and HARTMUT MICHEL
for the determination of the three-dimensional structure of a photosynthetic reaction center.

☆

Physiology or Medicine
SIR JAMES W. BLACK, GERTRUDE B. ELION and GEORGE H. HITCHINGS
for their discoveries of important principles for drug treatment.

☆

Literature
NAGUIB MAHFOUZ
who, through works rich in nuance-now clear-sightedly realistic, now evocatively ambiguous-has formed an Arabian narrative art that applies to all mankind.

☆

Peace
THE UNITED NATIONS PEACE-KEEPING FORCES
New York, NY, U.S.A.

☆

Economics
MAURICE ALLAIS
for his pioneering contributions to the theory of markets and efficient utilization of resources.

1988

SIR JAMES WHYTE BLACK (1925-2010),
GERTRUD BELLE ELION (1918-1999),
AND GEORGE HERBERT
HITCHINGS (1905-1998)

Sir James Whyte Black   **Gertrud Belle Elion**   **George Herbert Hitchings**

© The Nobel Foundation

James Black (left), **Gertrude Elion** (right), and **George Hitchings** jointly received the 1988 Nobel Prize in Physiology or Medicine, "for their discoveries of the important principles of drug treatment."

Born in Uddingston, Scotland, James Black studied medicine at the University of St. Andrews. In 1958, he joined the Pharmaceutical Division of Imperial Chemical Industries. In 1948, the American scientist Reymond Alquist had proposed that two sets of receptors were present—alpha and beta—that might explain the seemingly paradoxical actions of epinephrine and norepinephrine on the cardiac muscles. Black and his colleagues attempted to characterize these receptors. Using isoproterenol, an analogue of norepinephrine, they synthesized propranolol—a beta receptor antagonist, which became invaluable in the treatment of coronary artery diseases.

Black moved to Smith, Kline & French Company (SK& F; now Smith Kline Beecham) in 1964 and pursued antihistamine research. Since the antihistamines available then could inhibit nasal secretions, but not gastric acid secretions, Black proposed the existence of a different receptor (H_2), akin to the beta receptor. Using a series of histamine analogues, Black and coworkers developed several H_2 blockers, many of which were ineffective and toxic. By 1976, Black developed cimetidine—a powerful H_2 blocker useful in the treatment of gastric and peptic ulcers. Black had said that "the most fruitful basis for the discovery of a new drug is to start with an old drug."

Born in Hoquiam, Washington, Hitchings studied biochemistry at Harvard University, and in 1942 joined the Burroughs Wellcome (BW) company. In 1944, Gertrude Belle Elion, a New Yorker, with a master's degree in chemistry from the New York University, joined Hitchings and remained with him as collaborator for the rest of her career.

Elion and Hitchings' approach in pharmacology research was revolutionary. They discarded the old "magic bullet" method and applied the basic principles of biochemistry and physiology while studying drugs. Having found that bacteria needed folic acid and purines for DNA synthesis, they were able to develop 6-mercaptopurine (6 MP), a folic acid antagonist an effective chemotherapeutic agent against leukemia.

Applying the same principles that led them to 6 MP, Elion and Hitchings succeeded in producing a series of drugs. In 1950, they developed pyrimethamine; then came trimethoprim, azathioprine, and allopurinol; and in 1975, they synthesized acyclovir, a powerful antiviral agent against herpes virus. Elion and Hitchings' pioneering principles in pharmacology were also instrumental in the development of 5-fluorouracil, cytosine and adenine arabinosides, and more recently, azidothymidine (zidovudine, AZT).

Because of her sex, Elion faced numerous obstacles in her scientific career. A compassionate, inspiring, and industrious scientist—she never stopped working until her sudden death in February 1999—Elion once said, "The Nobel Prize is fine, but the drugs I have developed are rewards in themselves."

References

1. Ahlquist RP. Adrenergic Receptor: A Personal and Practical View. Perspectives in Biology and Medicine 17, pp. 119-121, Autumn 1973.
2. Black JW (Stephenson JS). Pharmacology of a New Adrenergic Beta-Receptor Blocking Compound (Nethalide). (with Crowther AF, Shanks RG, Smith LH, Dornhorst AC) A New Adrenergic Beta-Receptor Antagonist. Lancet, no. 2, pp. 311-314, 1962; (with Duncan WAM, Shanks RG) Comparison of Some Properties of Pronethalol and Propranolol. British Journal of Pharmacology, vol. 25, pp. 577-591, 1965.
3. Black JW (with Prichard BNC) Activation and Blockade of ß-Adrenoceptors in Common Cardiac Disorders. British Medical Bulletin, vol. 29, pp. 163-167, 1973; (with Duncan WAM, Emmett JC, Ganellin CR, Heselbo T, Parsons ME, Wyllie JH) Metiamide: An Orally Active Histamine H_2-Receptor Antagonist. Agents and Actions, vol. 3, pp. 133-137, 1973.
4. Burchall JJ, Hitchings GH. The Role of Metabolites and Antimetabolites in the Control of folate Coenzyme Synthesis. Advances in Enzyme Regulation 26, pp. 323-333, 1968.
5. Elion GB (with Hitchings G). Antagonists of Nucleic Acid Derivatives IV: Reversal Studies with 2-Amnopurinol and 126 Di aminopurine. Journal of Biological Chemistry, vol. 187, p. 511, 1950; (with Hitchings, Werff HV) Antagonists of Nucleic Acid Derivatives VI: Purines. Journal of Biological Chemistry, vol. 192, p. 505, 1951; (with Goodman I, Lange W, Hitchings GW) Condensed Pyrimidine Systems XX: Purines Related to 6-Meraptopurine and Thioguanine. Journal of American Chemical Society, vol. 81, pp. 1898-1902, 1959.
6. Elion GB. The Biochemistry and Mechanism of Action of Acyclovir. Journal of Antimicrobial Chemotherapy, vol. 12, pp. 9-17, 1983.

7. Hitchings GH (with Falco EA, Sherwood MB). The Effect of Pyrimidines on the Growth of Lactobacillus Caseii. Science, vol. 4, pp. 251-252, 1945; (with Elion GB, Falco EA, Russell PB, Vanderwerf H) Studies on Analogs of Purines and Pyrimidines. Annals of the New York Academy of Science, vol. 52, pp. 1318-1355, 1965; (with Burchall JJ) Inhibitor Binding Analysis of Dihydrofolate Reductases from Various Species. Molecular Pharmacology, vol. I, pp. 126-136, 1965.

1989

☆

Physics
NORMAN F. RAMSEY
for the invention of the separated oscillatory fields method and its use in the hydrogen maser and other atomic clocks, and
HANS G. DEHMELT and WOLFGANG PAUL
for the development of the ion trap technique.

☆

Chemistry
SIDNEY ALTMAN and THOMAS R. CECH
for their discovery of catalytic properties of RNA.

☆

Physiology or Medicine
J. MICHAEL BISHOP and HAROLD E. VARMUS
for their discovery of the cellular origin of retroviral oncogenes.

☆

Literature
CAMILO JOSÉ CELA
for a rich and intensive prose, which with restrained compassion forms a challenging vision of man's vulnerability.

☆

Peace
THE 14TH DALAI LAMA (TENZIN GYATSO)
Tibet. Religious and political leader of the Tibetan people.

☆

Economics
TRYGVE HAAVELMO
for his clarification of the probability theory foundations of econometrics and his analyses of simultaneous economic structures.

1989

John Michael Bishop   **Harold Eliot Varmus**

© The Nobel Foundation

John Michael Bishop (left), **and Harold Eliot Varmus** shared the Nobel Prize in Physiology or Medicine for 1989 for their discovery of the "cellular origin of retroviral oncogenes."

Michael Bishop was born in York, Pennsylvania. He developed a keen interest in research during medical studies at Harvard University. After graduation and the mandatory internship, he joined as a postdoctoral fellow at the National Institutes of Health (NIH). In 1968, he moved to the University of California, San Francisco (UCSF).

Harold Varmus was born in Oceanside, New York, and studied medicine at Columbia University's medical school. Like Bishop, Varmus also joined the NIH as a Clinical Associate, where he got his "first serious exposure to laboratory science." In 1970, he joined Michael Bishop at the UCSF. Over the next two decades, the pair collaborated on retrovirus research.

In the early 1910s, Peyton Rous had shown that cell-free filtrates from mouse tumors, when injected into healthy chickens, caused sarcomas. The specific "filterable agent" was later identified as the Rous Sarcoma virus (RSV), belonging to the large family of retroviruses.

In the 1960s, the term "oncogene" was proposed to denote the genetic elements of the virus responsible for inducing tumor in the host. However, with the knowledge of the newly discovered viral reverse transcriptase enzyme, and using variants of the RSV, Bishop and Varmus developed DNA probes to study the mechanism of tumor genesis. In 1975, they made the startling discovery that the oncogene was not a true genetic agent from the virus. Instead, it was part of the normal host gene, which the virus had acquired during replication and carried along thereafter. Later it was shown that the oncogenes play a critical role in growth and division of normal cells.

The discovery of the nature of oncogenes had a profound effect in understanding tumor pathogenesis. Oncogenes with point mutations have been seen in several tumors, including Burkitt's lymphoma, neuroblastoma, and breast cancer.

A self-confessed book addict, Bishop said that high school tests predicted him to become a journalist, forestry expert, or music teacher. "If offered reincarnation," he added, "I would choose the career of a performing musician... preferably in a string quartet."

In 1993, President Clinton named Varmus as the head of the NIH, where Varmus was responsible for major reforms. He single-handedly succeeded in persuading Congress to invest heavily in biomedical research—the result was a huge increase in the NIH budget. In December 1999, he left the NIH to head the Memorial Sloan-Kettering Cancer Center in New York. An avid outdoorsman, Varmus may have trouble bicycling to work in Manhattan as he did for six years at the NIH.

References

1. Baltimore D. Viral RNA-Dependent DNA Polymerase. Nature 226, pp. 1209-1211, 1970.

2. Bishop JM (with Varmus HE, Weiss RA, Friis RR, Levinson W). Detection of Avian Tumor Virus Specific Nucleotide Sequences in Avian Cell DNAs. Proceedings of the National Academy of Sciences, USA, vol. 69, pp. 20-24, 1972; (with Varmus HE, Heasley S) Use of DNA-DNA Annealing to Detect New Virus-Specific DNA Sequences in Chicken Embryo Fibroblasts After Infection by Avian Sarcoma Virus. Journal of Virology, vol 14, pp. 895-903, 1974.

3. Bishop JM. The Molecular Genetics of Cancer. Science 235, pp. 305-307, January 16, 1987.

4. Varmus HE (with Weiss RA, Friis RA, Levinson W, Bishop JM). Detection of Avian Tumor Virus Specific Nucleotide Sequences in Avian Cell DNAs. Proceedings of the National Academy of Sciences, USA, vol. 69, pp. 20-24, 1972; (Vogt PK, Bishop) Integration of Deoxyribonucleic Acid Specific for Rous Sarcoma Virus After Infection of Permissive and Nonpermissive Host. Proceedings of the National Academy of Sciences, USA, vol. 70, pp. 3067-3071, 1973; (with Stehelin D, Bishop, Vogt) DNA Related to the Transforming Genes) of Avian Sarcoma Viruses Is Present in Normal Avian Cells. Nature, vol. 260, pp. 170-173, 1976.

5. Varmus HE. Oncogenes and Transcriptional Control. Science, vol. 238, pp. 1337-1339, 1987; Reverse Transcription. Scientific American, vol. 257, p. 56, 1987; Lessons from the Life Cycle of Retroviruses. The Harvey Lectures 1987-1988, vol. 83, pp. 35-56 1989.

1990

☆

Physics
JEROME I. FRIEDMAN, HENRY W. KENDALL
and RICHARD E. TAYLOR

for their pioneering investigations concerning deep inelastic scattering of electrons on protons and bound neutrons, which have been of essential importance for the development of the quark model in particle physics.

☆

Chemistry
ELIAS JAMES COREY

for his development of the theory and methodology of organic synthesis.

☆

Physiology or Medicine
JOSEPH E. MURRAY and E. DONNALL THOMAS

for their discoveries concerning organ and cell transplantation in the treatment of human disease.

☆

Literature
OCTAVIO PAZ

for impassioned writing with wide horizons, characterized by sensuous intelligence and humanistic integrity.

☆

Peace
MIKHAIL SERGEYEVICH GORBACHEV

President of the USSR, helped to bring the Cold War to an end.

☆

Economics
HARRY M. MARKOWITZ, MERTON M.
MILLER and WILLIAM F. SHARP

for their pioneering work in the theory of financial economics.

1990

JOSEPH EDWARD MURRAY (1919-2012), AND E. DONNALL THOMAS (1920-2012).

Joseph Edward Murray   **Edward Donnall Thomas**
© The Nobel Foundation

Joseph Murray (left), and **Donnall Thomas** shared the 1990 Nobel Prize in Physiology or Medicine for discoveries concerning "organ and cell transplantation in the treatment of human disease."

Joseph Murray was born in Milford, Massachusetts and studied medicine at Harvard University. After internship, he served in the US military, where he became adept at treating battlefield injuries, including skin grafting and plastic surgery methods. Then he joined the Children's Hospital Medical Center in Boston and the surgical faculty at Harvard Medical School.

The concept of organ transplantation had been around for a long time. According to the Christian legends, Saints Damian and Cosmos amputated the ulcerated leg of a Caucasian and replaced it with that of a dead Moor. During the early decades of 20th century, some surgeons grafted tissues successfully in experimental animals, but tissue rejection remained a major hurdle.

Murray too began work in animals. Having understood the evolving concepts of immunological tolerance, Murray perfected transplantation techniques and performed a historic procedure on December 23, 1954. In a patient with terminal renal failure, Murray and colleagues transplanted a kidney donated by the patient's healthy, identical twin. The procedure was a technical marvel and the new kidney in the patient began working almost instantly; the donor, too, remained well.

By using the immunosuppressive drugs introduced in the 1950s and 1960s, Murray made history again in 1962. He showed that kidneys from donors unrelated to the recipients can also be transplanted successfully, if tissue-typing was carried out. Those concepts were soon extended to organs obtained from unrelated cadavers, heralding the modern era of organ transplantation.

Son of a doctor, E. Donnall Thomas was born in Mart, Texas and studied medicine at Harvard Medical School. A colleague of Murray in Boston, Thomas studied hematology. In 1963, he joined the University of Washington School of Medicine in Seattle.

It was known that transfused marrow cells functioned well in the recipient's body for brief periods. But attempts to maintain the donor cells functioning in the recipients for prolonged periods had failed, either because of rejection or because of the devastating graft-versus-host disease (GVHD). With Joseph Ferrebee and other collaborators, in 1955 Thomas developed methods to store and transfuse large quantities of bone marrow in animals. However, his early work on marrow transplantation in patients with leukemia failed, despite prior irradiation of the recipients to get rid of the malignant cells.

In 1969, Thomas and colleagues made a breakthrough. After HLA typing and by using immune suppressive therapy and total body irradiation, they performed a bone marrow transplantation on a 46 – year-old man with nearly terminal chronic granulocytic leukemia. The donor was his sister, whose HLA typing was 'reasonably compatible.' On March 10, they infused 360 ml of bone marrow, obtained from the donor. The patient showed clear evidence of recovery from 15-41 days after infusion in blood counts. Fifty-six days later, however, the patient died of a 'disseminated viral disease, probably cytomegalovirus,' as per the autopsy. It would take many more trials before a consistently good outcome could be documented in patients receiving bone marrow transplantation.

Like the kidney, bone marrow transplantations too began to receive worldwide acceptance during the subsequent decades. Now these procedures are performed on a routine basis.

References

1. Buckner CD, Epstein RB, et al. Allogeneic marrow engraftment following whole body irradiation in a patient with leukemia. Blood, Volume 35; 741—50, 1970.

2. Murray JE, Thomas ED (with Merrill JP, Harrison JH, Guild WR) Successful transplantation of the Human Kidney Between Identical Twins. Journal of the American Medical Association, vol. 160, pp. 277-282, 1956; (with Merrill JP, Harrison JH) Kidney Transplantation Between Seven Pairs of Identical Twins. Annals of Surgery, vol. 148, pp. 343-359, 1958.

3. Murray JE, Thomas ED (with Merrill JP, Takacs F, et al). Successful Transplantation of Kidney from a Human Cadaver. Journal of the American Medical Association, vol. 185, pp. 347-353, 1963.

4. Thomas ED (with Lochte HL, Lu WC, et al) Intravenous Infusion of Bone Marrow in Patients Receiving Radiation and Chemotherapy. New England Journal of Medicine, vol. 257, pp. 491-496, 1957; (with Lochte HL, Cannon JH, Sahler OD, Ferrebee JW) Supralethal Whole Body Irradiation and Isologous Marrow Transplantation in Man. Journal of Clinical Investigation, vol. 38, pp. 1709-1716, 1959.

5. Thomas ED (with Storb R, Clift RA, et al). Bone Marrow Transplantation. New England Journal of Medicine, vol. 292, pp. 832-843, 1975; Storb R, Weiden PL, et al) Aplastic Anemia Treated by Allogeneic Bone-Marrow Transplantation: A Report on Forty-nine New Cases from Seattle. Blood, vol. 48, pp. 817-841, 1976.

6. Thomas ED, Buckner CD, and others: Allogeneic marrow grafting for hematologic malignancy using HL-A matched donor-recipient sibling pairs. Blood. Vol 38; 267—87, 1971.

1991

☆

Physics
PIERRE-GILLES DE GENNES

for discovering that methods developed for studying order phenomena in simple systems can be generalized to more complex forms of matter, in particular to liquid crystals and polymers.

☆

Chemistry
RICHARD R. ERNST

for his contributions to the development of the methodology of high resolution nuclear magnetic resonance (NMR) spectroscopy.

☆

Physiology or Medicine
ERWIN NEHER and BERT SAKMANN

for their discoveries concerning the function of single ion channels in cells.

☆

Literature
NADINE GORDIMER

who through her magnificent epic writing has - in the words of Alfred Nobel – "been of very great benefit to humanity."

☆

Peace
AUNG SAN SUU KYI

Burma. Oppositional leader, human rights advocate.

☆

Economics
RONALD H. COASE

for his discovery and clarification of the significance of transaction costs and property rights for the institutional structure and functioning of the economy.

1991

ERWIN NEHER (BORN 1944)
AND BERT SAKMANN (1942)

Erwin Neher   **Bert Sakmann**
© The Nobel Foundation

Erwin Neher (left), and **Bert Sakmann** won the Nobel Prize in Physiology or Medicine for 1991 "for their discoveries concerning the function of single ion channels in cells."

Neher was born in Landsberg, Germany and grew up in Buchloe, a small Bavarian town. Developing an interest in 'living things' and in technical and analytical aspects of physics and mathematics while in high school, he had decided to study biophysics. At Munich's Technical University he earned his B.S., and later a Ph.D. In 1966 he went to the University of Wisconsin in Madison, on a Fulbright Fellowship, and earned his M.S., degree. In 1967 he returned to Munich, joined the Max-Plank Institute for Psychiatry and pursued research on ion channels and on synaptic transmission.

Sakmann was born in Stuttgart, Germany. Interested in physics and engineering since childhood, he would design and build model ships and remote-control airplanes. He studied medicine at Tübingen University, and in 1970 went to London's University College as postdoctoral fellow under Professor Bernard Katz. There he studied the basics of neurotransmitter research and techniques of biophysical measurements. In 1973, he received a position in the Max-Plank Institute for Biochemical Chemistry in Göttingen, where Neher also had moved; the two were to collaborate for 16 years.

It had long been known that ionic chemicals entered and exited cell membranes freely, but the routes of transports were difficult to determine because of the extremely small ionic currents involved in channel trafficking. By applying molecular biology techniques to the problems of ion physiology, Neher and Sakmann studied channel conductance.

They developed a thin glass micropipette as a recording electrode. When brought in contact with the cell membrane, it formed a tight seal with the periphery of the pipette orifice. Therefore, the exchange of ions between the inside of the pipette and the outside could only occur through the ion channel in the membrane fragment. When a single ion channel opened, ions moved through the channels as electric current. Through a refinement of the electronic equipment and recording conditions Neher and Sakmann succeeded in measuring miniscule amounts of current.

This 'patch-clamp' technique was soon found to have widespread research applications. It helped identify ion transfer channels across cell membranes and factors that control them. It aided in the discovery of the mechanisms of actions of cyclic AMP, histamine, insulin, epinephrine and other hormones, and in finding molecular abnormalities in cystic fibrosis (chloride ion channels), epilepsy (sodium and potassium ion channels), and Lambert-Eatons disease (calcium ion channels). Today patch-clamp technique is one of the most widely used research tools in molecular biology and applied pharmacology.

References

1. Alberts et al.: The Molecular Biology of the Cell. Garland Press, 1990, 2nd edition, pp. 156, 312-326, 1065-1084.
2. Grillner, S. I: N. Calder (ed.). Scientific Europe. Foundation Scientific Europe, 1990.
3. Gurney AM. Electrophysiological recording methods used in vascular biology. J Pharmacol Toxicol Methods. 2000 Sep-Oct;44(2):409-20.
4. Sucher NJ, Deitcher DL, Baro DJ, Warrick RM, Guenther E. Genes and channels: patch/voltage-clamp analysis and single-cell RT-PCR. Cell Tissue Res. 2000 Dec;302(3):295-307.

1992

☆
Physics
GEORGES CHARPAK
for his invention and development of particle detectors, in particular the multiwire proportional chamber.

☆
Chemistry
RUDOLPH A. MARCUS
his contributions to the theory of electron transfer reactions in chemical systems.

☆
Physiology or Medicine
EDMOND H. FISCHER and EDWIN G. KREBS
for their discoveries concerning reversible protein phosphorylation as a biological regulatory mechanism.

☆
Literature
DEREK WALCOTT
for a poetic oeuvre of great luminosity, sustained by a historical vision, the outcome of a multicultural commitment.

☆
Peace
RIGOBERTA MENCHU TUM
Guatemala. Campaigner for human rights, especially for indigenous peoples.

☆
Economics
GARY S. BECKER
for having extended the domain of microeconomic analysis to a wide range of human behavior and interaction, including non-market behavior.

1992

EDMOND HENRI FISCHER (1920-2021)
AND EDWIN GERHARD
KREBS (1918-2009)

Edmond Henri Fischer   Edwin Gerhard Krebs
© The Nobel Foundation

Edmond Fischer (left), and **Edwin Krebs** won the 1992 Nobel Prize in Physiology or Medicine for their discoveries concerning "reversible protein phosphorylation as a biological regulatory mechanism."

Edwin Krebs was born in Lansing, Iowa. Because his father was a Presbyterian minister and mother, a high school teacher, the family moved frequently. Edwin studied biochemistry at the University of Illinois, Urbana and medicine at Washington University, St. Louis. After serving briefly in the US Navy, he returned to his alma mater in St. Louis but could not secure a clinical post right away. Hence, he pursued his old love, biochemistry research, joining Carl and Gerty Cori. In 1948, Krebs moved to Seattle's University of Washington as assistant professor of biochemistry.

Son of an Austrian father and French mother, Edmond Fischer was born in Shanghai, China. While studying in the all-boys' Collège de Calvin in Geneva, Switzerland, he and a classmate of his decided to study science and medicine, so that together they could "cure all the ills of the world." After obtaining a degree from the School of Chemistry, Fischer went to the USA and joined Krebs in Seattle, at the University of Washington. The Fisher couple loved the picturesque mountains, forests and lakes in Seattle so much that they decided to settle there, since it reminded them of their beautiful Switzerland.

At that time, Earl Sutherland had worked on liver phosphorylase. Having discovered cyclic AMP he had shown that it was the second messenger of hormone actions and was involved in phosphorylase *a* formation. Fischer's expertise in potato phosphorylase research and Krebs' expertise in mammalian skeletal muscle work complemented their combined research on glycogen breakdown pathways. By the mid-1950s they had discovered that phosphorylase can be converted from an inactive to an active form by the transfer of a phosphate group from ATP to the protein and the process was catalyzed by protein kinase—the first enzyme to be discovered in 'reverse' phosphorylation.

Several years elapsed before it became apparent that reversible protein phosphorylation was a general process affecting countless cellular proteins. It was involved in such diverse phenomenon as mobilization of glucose from glycogen, prevention of transplant rejection (by cyclosporin), and the development of chronic myeloid leukemia. Later work also showed that approximately 1% of the genes in the genome encoded for the protein kinases in the body.

Besides their respective subjects, these scientists cultivated diverse interests in arts and literature. An avid outdoorsman, Krebs is a voracious reader. Classical music has been an integral part of Fischer's life. Having studied at the Geneva Conservatory of Music, it seems that once he had even considered a career in music.

References

1. Alberts et al. The Molecular Biology of the Cell. Garland Press, 1990, 2nd edition. Phosphorylation-dephosphorylation cycle of proteins pp. 129-131, 710-712, 736-737, 777-778.

1993

☆

Physics
RUSSELL A. HULSE and JOSEPH H. TAYLOR JR.

for the discovery of a new type of pulsar, a discovery that has opened up new possibilities for the study of gravitation.

☆

Chemistry
KARY B. MULLIS

for his invention of the polymerase chain reaction (PCR) method, and **MICHAEL SMITH**

for his fundamental contributions to the establishment of oligonucleotide-based, site-directed mutagenesis and its development for protein studies.

☆

Physiology or Medicine
RICHARD J. ROBERTS and PHILLIP A. SHARP

for their independent discoveries of split genes.

☆

Literature
TONI MORRISON

who in novels characterized by visionary force and poetic import, gives life to an essential aspect of American reality.

☆

Peace
NELSON MANDELA
Leader of the ANC., and
FREDRIK WILLEM DE KLERK
President of the Republic of South Africa.

☆

Economics
ROBERT W. FOGEL and DOUGLASS C. NORTH

for having renewed research in economic history by applying economic theory and quantitative methods in order to explain economic and institutional change.

1993

RICHARD JOHN ROBERTS (BORN 1943)
PHILLIP A. SHARP (BORN 1944)

Sir Richard John Roberts   **Phillip Allen Sharp**

© The Nobel Foundation

Richard John Roberts (left), and **Phillip A. Sharp** shared the 1993 Nobel Prize for Physiology or Medicine "for their discovery of split genes."

Richard Roberts was born in Derby, England and studied chemistry in Sheffield University earning his Ph.D. In 1969 he moved to Harvard University as postdoctoral fellow. Three years later he accepted James Watson's invitation and joined Cold Spring Harbor Laboratory (CSHL), Long Island. In 1992 he moved to New England Biolabs, a small biotechnology company producing research chemicals and restriction enzymes.

Sharp was born in a rural community in the northern hill country of Kentucky. Earning money for college tuition from raising cattle and growing tobacco, he studied at Union College in eastern Kentucky and later received his Ph.D., in chemistry from the University of Illinois. He was greatly influenced by *The Genetic Code,* a 1966 monograph from a symposium held at the CSHL, and decided to study molecular biology. Sharp studied and worked at many institutions, including CSHL, California Institute of Technology, and the Massachusetts Institute of Technology (MIT).

Ever since the discovery of the molecular structure of DNA, biologists had believed that genes were made of uninterrupted strands of the genetic material. Working with the adenovirus, in 1977, the teams of Roberts and Sharp independently discovered a feature of gene structure that dramatically changed molecular biology. These scientists were studying the relationship between the genome and its mRNA. They created hybrid molecules in which mRNA strands bind to their complementary DNAs. Surprisingly, the hybrid mRNAs did not match the DNA—when examined under the electron microscope, the scientists found that some segments of the DNA were ignored in making the final RNA, and hence the protein.

From this, the scientists concluded that instead of being continuous, the gene materials contained several coding and non-coding segments. These segments were later named 'exons' and 'introns' by Harvard University's Walter Gilbert. It was soon shown that discontinuous, or 'split' structure of the gene was a common genetic feature.

Roberts' and Sharp's discovery led to the understanding of 'gene splicing,' an essential feature in expressing genetic information. The concept of introns and exons and gene splicing became the basis for a strikingly new explanation of evolution. It has also helped identify the molecular defects in such diverse conditions as beta-thalassemia and chronic myeloid leukemia.

Roberts has described himself as "a passionate reader." It seems that until he discovered chemistry, he had wanted to become a detective, because "it seemed that they paid you to solve puzzles."

References

1. Alberts B (ed) Molecular Biology of the Cell. Garland, New York, 1989
2. Chambon P: Split Genes. Scientific American 244, 60-71 (1981)
3. Darnell JE et al: Molecular Cell Biology. Scientific American Books. Freeman, New York, 1990
4. Darnell JE: RNA. Scientific American 253, 68-78 (1985)
5. Greenspan RJ. The flexible genome. Nat Rev Genet. 2001 May; 2(5):383-7.
6. McConkey EH: Human Genetics. The Molecular Revolution. Jones and Bartlett, Boston, 1993.
7. Ochman H, Moran NA. Genes lost and genes found: evolution of bacterial pathogenesis and symbiosis. Science. 2001 May 11;292(5519):1096-9.
8. Ronen A, Glickman BW. Human DNA repair genes. Environ Mol Mutagen. 2001;37(3):241-83.
9. Schofield PN, Joyce JA, Lam WK, et. al. Genomic imprinting and cancer; new paradigms in the genetics of neoplasia. Toxicol Lett. 2001 Mar 31;120(1-3):151-60.
10. Woodgate R. Evolution of the two-step model for UV-mutagenesis. Mutat Res. 2001 Feb 25;485(1):83-92.

1994

☆

Physics
BERTRAM N. BROCKHOUSE
for the development of neutron spectroscopy
CLIFFORD G. SHULL
for the development of the neutron diffraction technique.

☆

Chemistry
GEORGE A. OLAH
for his contribution to carbonation chemistry.

☆

Physiology or Medicine
ALFRED G. GILMAN and MARTIN RODBELL
*for their discovery of G-proteins and the role of these
proteins in signal transduction in cells.*

☆

Literature
KENZABURO OE
*who with poetic force creates an imagined world, where life and myth
condense to form a disconcerting picture of the human predicament today.*

☆

Peace
YASSER ARAFAT
Chairman of the Executive Committee of the PLO and
President of the Palestinian National Authority
SHIMON PERES
Foreign Minister of Israel, and
YITZHAK RABIN
Prime Minister of Israel
for their efforts to create peace in the Middle East.

☆

Economics
JOHN C. HARSANYI, JOHN F. NASH and REINHARD SELTEN

for their pioneering analysis of equilibria in the
theory of non-cooperative games.

1994

ALFRED GOODMAN GILMAN (1941-2015), AND MARTIN RODBELL (1925-1998)

Alfred Goodman Gilman   **Martin Rodbell**

© The Nobel Foundation

Alfred G. Gilman (left), **and Martin Rodbell** were awarded the 1994 Nobel Prize in Physiology or Medicine "for their discoveries of G-proteins and the role of these proteins in signal transduction in cells."

Martin Rodbell was born in Baltimore, Maryland and studied medicine at Johns Hopkins University. After earning a Ph.D. in biochemistry from the University of Washington in Seattle, he studied as a post-doctoral fellow at the University of Illinois at Urbana, Illinois and since 1956, worked at the National Institutes of Health (NIH) in Bethesda.

Alfred Gilman's father (also Alfred Gilman), the renowned pharmacologist, 'played almost any musical instrument' and his mother, a superb pianist, gave music lessons. Born in New Haven, Connecticut, Gilman also developed an interest in music and in science—the latter from visits to his father's pharmacology laboratory. He studied biochemistry at Yale University, and earned his Ph.D., from Case Western Reserve University in Cleveland before joining the NIH. Gilman later served as chairman of the Department of Pharmacology at the University of Texas in Dallas.

For a long time, the precise mechanism by which chemical signals were transmitted from cell exterior to cell interior (transduction) was unknown. In a series of pioneering research in the late 1960s and the early 1970s, Rodbell and coworkers showed that message transduction entailed several key steps. An extracellular signal (or the first messenger, e.g., a molecule of adrenaline) is recognized by a specific receptor, or "discriminator," activating a cell-surface switch, the "transducer," which requires guanosine 5'-triphosphate (GTP). This, in turn, stimulates an "amplifier," or the "second messenger" (e.g., cyclic AMP) for executing the actions contained in the signal.

Gilman and coworkers explored the chemical nature of Rodbell's transducer and found that genetically mutated leukemia cells fail to respond to external signals. By 1977, they had identified a protein lost during mutation as the cause for such a loss of function. They also showed that the transducer function can be restored by supplementing the protein onto the cell membrane. In 1980, the chemical was shown to be the G-protein, so named because it binds with GTP.

G-proteins were soon found to be one of the most essential molecular switches, composed of three peptide chains of varying length. A conservative estimate has it that at least a third of all signal transmission involves G-proteins. These proteins mediate proper functioning of vision, smell, and taste. Dysfunctions of G-protein expression and activity explain symptoms of such varied disorders as cholera, pertussis, diabetes, and alcoholism. Many genetic conditions (e.g., the McCune Albright syndrome) have been found to be due to G-protein-related abnormalities.

The middle name, 'Goodman' was given to Gilman because his father, Alfred Gilman and Louis S. Goodman published the first edition of their famous textbook *The Pharmacological Basis of Therapeutics* in

1941—the year of Alfred's birth. Alfred Goodman Gilman may be the only Nobel laureate named after a medical textbook.

References

1. Brown EM. G protein-coupled, extracellular Ca2+ (Ca2+(o))-sensing receptor enables Ca2+(o) to function as a versatile extracellular first messenger. Cell Biochem Biophys. 2000;33(1):63-95.
2. Chen CA, Manning DR. Regulation of G proteins by covalent modification. Oncogene. 2001 Mar 26;20(13):1643-52.
3. Cismowski MJ, Takesono A, Bernard ML, Duzic E, Lanier SM. Receptor-independent activators of heterotrimeric G-proteins. Life Sci. 2001;68(19-20):2301-8.
4. Radhika V, Dhanasekaran N. Transforming G proteins. Oncogene. 2001 Mar 26;20(13):1607-14.
5. Ram PT, Iyengar R. G protein coupled receptor signaling through the Src and Stat3 pathway: role in proliferation and transformation. Oncogene. 2001 Mar 26;20(13):1601-6.
6. Yokomizo T, Izumi T, Shimizu T. Leukotriene B4: metabolism and signal transduction. Arch Biochem Biophys. 2001 Jan 15;385(2):231-41.

1995

☆

Physics
MARTIN L. PERL
for the discovery of the tau lepton and
Frederick Reines
for the detection of the neutrino.

☆

Chemistry
PAUL CRUTZEN, MARIO MOLINA, and F. SHERWOOD ROWLAND
for their work in atmospheric chemistry, particularly concerning the formation and decomposition of ozone.

☆

Physiology or Medicine
EDWARD B. LEWIS, CHRISTIANE NÜSSLEIN-VOLHARD and ERIC F. WIESCHAUS
for their discoveries concerning the genetic control of early embryonic development.

☆

Literature
SEAMUS HEANEY
for works of lyrical beauty and ethical depth, which exalt everyday miracles and the living past.

☆

Peace
JOSEPH ROTBLAT and THE CONFERENCES ON SCIENCE AND WORLD AFFAIRS
for their efforts to diminish the part played by nuclear arms in international politics and in the longer run to eliminate such arms.

☆

Economics
ROBERT LUCAS

for having developed and applied the hypothesis of rational expectations and thereby having transformed macroeconomic analysis and deepened our understanding of economic policy.

1995

EDWARD BUTTS LEWIS (1918-2004), CHRISTIANE NÜSSLEIN-VOLHARD (BORN1942), AND ERIC FRANCIS WIESCHAUS (BORN 1947)

Edward Butts Lewis   **Christiane Nüsslein-Volhard**   **Eric Francis Wieschaus**

© The Nobel Foundation

Edward Lewis (left), **Christiane Nüsslein-Volhard** (middle), and **Eric Wieschaus** shared the 1995 Nobel Prize in Physiology or Medicine "for their discoveries concerning the genes that control early embryonic development."

Edward Lewis was born in Wilkes-Barre, Pennsylvania and studied at the University of Minnesota. After earning his Ph.D., from the California Institute of Technology (Caltech) in 1942, he served in the US Air Force as captain, meteorologist and oceanographer. He rejoined Caltech in 1946 and worked there for much of his professional career.

Eric Wieschaus was born in South Bend, Indiana and earned his Ph.D. from Yale University. Before joining Princeton University in 1981, he worked for three years as Group Leader at the European Molecular

Biology Laboratory (EMBL) in Heidelberg, Germany, collaborating with Christiane Nüsslein-Volhard in genetic research.

Born in Madgeburg, Germany, Christiane Nüsslein-Volhard obtained her Ph.D., from the University of Tübingen. She worked in laboratories at Basel, Freiburg, and the EMBL at Heidelberg, before returning to Tübingen's Max Planck Institutes.

In the early 1940s, Lewis began studying genetics using the fruit fly *Drosophila*. He explored the genetic basis for a strange mutation that caused "monster" flies with four instead of two wings. Lewis found that the extra pair of wings occurred from duplication of an entire body segment. Studying specific genes that might control specialization of larval body segments, Lewis discovered the "co-linearity principle" according to which gene location specified differentiation along the anterior-posterior body axis. Genes at the beginning of the complex controlled anterior body segments while those further down controlled posterior segments.

Lewis' summary of his life-long work in the late 1970s had an enormous impact on the future of genetic research and was a source of inspiration for Nüsslein-Volhard and Wieschaus. To discover the genes responsible for body segmentation in the developing fly larvae, Nüsslein-Volhard and Wieschaus treated female flies with chemicals to induce random gene mutations and studied body axis and segmentation patterns in the offspring. With tenacity, hard work and perseverance, they examined some 40,000 mutations and identified 15 genes controlling the early phases of embryonic development in the *Drosophila*. They presented these findings in their landmark 1980 report in *Nature*.

The genes identified by Nüsslein-Volhard and Wieschaus also play a critical role during the development of the human embryo. They influence formation of the body axis and body segmentation, and specialization of individual segments into specific organs. About 40% of idiopathic congenital malformations may be due to mutations in genes discovered by Nüsslein-Volhard and Wieschaus.

About the contributions of these scientists, it has been said that if their work were removed from the textbooks of genetics, very little would be left.

References

1. Alberts, B. et al (Eds). The Molecular Biology of the Cell. 3rd edition pp 1077-1107. Garland Publishing, New York 1994.
2. Lawrence, P. The Making of a Fly. Blackwell Scientific Publications. Oxford 1992.
3. Lewis, E.B. (1978) A Gene Complex Controlling Segmentation in Drosophila. Nature 276, 565-570
4. McGinnis, W., Kuziora, M. (1994). The Molecular Architects of Body Design. Scientific American 270, 36-42
5. Nüsslein-Volhard, C., Wieschaus, E. (1980). Mutations Affecting Segment Number and Polarity in Drosophila. Nature 287, 795-801

1996

☆

Physics
**DAVID M. LEE, DOUGLAS D. OSHEROFF
and ROBERT C. RICHARDSON**
for their discovery of superfluidity in helium-3.

☆

Chemistry
**ROBERT F. CURL, Jr., SIR HAROLD W.
KROTO, and RICHARD E. SMALLEY**
for their discovery of fullerenes.

☆

Physiology or Medicine
PETER C. DOHERTY and ROLF M. ZINKERNAGEL
*for their discoveries concerning the specificity of
the cell mediated immune defense.*

☆

Literature
WISLAWA SZYMBORSKA
*for poetry that with ironic precision allows the historical and
biological context to come to light in fragments of human reality.*

☆

Peace
CARLOS FELIPE XIMENES BELO and JOSE RAMOS-HORTA
*for their work towards a just and peaceful
solution to the conflict in East Timor.*

☆

Economics
JAMES A. MIRRLEES and WILLIAM VICKREY
*for their fundamental contributions to the economic theory
of incentives under asymmetric information.*

1996

Peter Charles Doherty   **Rolf Martin Zinkernagel**
© The Nobel Foundation

Peter Doherty (left), **and Rolf Zinkernagel** shared the 1996 Nobel Prize in Physiology or Medicine "for the discovery of the specificity of the cell mediated immune defense."

Raised in the outskirts of Brisbane, Australian Peter Doherty studied veterinary science. In 1967, he went to the Moredun Research Institute in Edinburgh as an experimental pathologist and earned his Ph.D. from Edinburgh University in 1970. The same year he returned to Australia and joined the John Curtin School of Medical Research (JCSMR) in Canberra. Later, he worked at many institutions, including St. Jude Children's Research Hospital and the University of Tennessee, both in Memphis.

Born in Basel, Switzerland, Rolf Zunkernagel studied medicine there. In 1973, he went to JCSMR as a visiting fellow, where he and Doherty were "thrown together" by chance. Because of lack of space, they had to share the same lab, whereupon they began research collaboration.

They tried to find why mice infected with lymphocytic chorio-meningitis virus (LCMV) developed lethal brain lesions. Their working hypothesis was that the killer T-lymphocytes produced in response to LCMV infection somehow caused brain injury; to test this, they developed a simple mouse model.

Having isolated the specific anti-LCMV T lymphocytes, they discovered that the lymphocytes recognized both the LCMV antigen and the major histocompatibility antigen in mice (like the HLA-antigen in man). It turned out that without the latter, the T cell immunity was ineffective. In explaining this unexpected finding, Doherty and Zunkernagel proposed the principle of "simultaneous dual recognition," according to which the immune system must recognize "self" and "non-self" simultaneously for long-term immunological efficacy.

These findings published in 1974 (*Nature* 248,70-2) had immediate and widespread impact on immunology research. It was shown that the set of T-lymphocytes that matured and survived in the human body did do so by their ability to recognize both the foreign and the HLA antigens simultaneously. Later, the molecular and structural basis for the dual recognition principle was discovered. The basic immunological principle discovered by Doherty and Zunkernagel became the foundation for developing vaccines against infectious diseases and more recently, against metastatic cancers.

After working for several years in the USA, Zunkernagel returned to Switzerland and became the head of the Institute of Experimental Immunology in Zurich in 1992.

When Doherty was in high school, he was very much influenced by the works of Huxley, Sartre and Hemingway. Calling himself a non-conformist marching to 'his own drum,' he bemoans that there are "too few people working on viral immunology, too little funding, too many problems, and too little time."

References

1. Doherty PC, Zinkernagel RM. A biological role for the major histocompatibility antigens. Lancet, 1406-1409, 1975.
2. Zinkernagel RM, Doherty PC. Immunological surveillance against altered self-components by sensitized T lymphocytes in lymphocytic choriomeningitis. Nature 251, 547-548, 1974.
3. Zinkernagel RM, Doherty PC. MHC restricted cytotoxic T cells: Studies on the biological role of polymorphic major transplantation antigens determining T cell restriction specificity. Advances in Immunology 27, 51-177, 1979
4. Zinkernagel RM, Doherty PC. Restriction of in vitro T cell-mediated cytotoxicity in lymphocytic choriomeningitis within a syngeneic and semi allogeneic system. Nature 248, 701- 702, 1974.

1997

☆

Physics
STEVEN CHU, CLAUDE COHEN-TANNOUDJI and WILLIAM D. PHILLIPS
for development of methods to cool and trap atoms with laser light.

☆

Chemistry
PAUL D. BOYER and JOHN E. WALKER
for their elucidation of the enzymatic mechanism underlying the synthesis of adenosine triphosphate (ATP) and
JENS C. SKOU
for the first discovery of an ion-transporting enzyme, Na+, K+-ATPase.

☆

Physiology or Medicine
STANLEY B. PRUSINER
for his discovery of Prions - a new biological principle of infection

☆

Literature
DARIO FO
who emulates the jesters of the Middle Ages in scourging authority and upholding the dignity of the downtrodden.

☆

Peace
INTERNATIONAL CAMPAIGN TO BAN LANDMINES (ICBL) and JODY WILLIAMS
for their work for the banning and clearing of anti-personnel mines.

☆

Economics
ROBERT C. MERTON and MYRON S. SCHOLES
for a new method to determine the value of derivatives.

1997

STANLEY BEN PRUSINER (BORN 1942)

© The Nobel Foundation

Stanley Ben Prusiner was awarded the Nobel Prize "for his discovery of prions."

He was born in Des Moines, Iowa and received his M.D. from the University of Pennsylvania in Philadelphia. After training in neurology, he joined the University of California in San Francisco, where he became professor of neurology, virology and biochemistry in 1984.

Prusiner's interest in the devastating, Creutzfeldt-Jakob disease (CJD) was triggered by the death of a patient of his with CJD in 1972. It was known then that CJD was like kuru seen among the Fore-people of New Guinea, and to scrapie seen in sheep. Extracts of the affected brain from patients with these diseases had been shown to transmit the disease to experimental animals. In the absence of a specific infectious etiology, it had been hypothesized that these disorders were due to "slow virus" infection.

Having read 'everything that was written on the subject,' in 1974 Prusiner tried to isolate the elusive etiological agent of scrapies with a grant from the National Institutes of Health. After years of hard work, he and his colleagues extracted a preparation from scrapies-infected hamster brain that had the causative agent. Prusiner called the new infectious particles, "prions," for "proteinaceous infectious particles."

His 1982 report in science on prions set off a firestorm of controversy, because the agent was neither a virus nor an "organism," and contained no DNA or RNA. Two years later, Prusiner and colleagues isolated a gene probe and showed that prion genes were present in all animals, including man. Prion proteins (PrP) were shown to be folded into two distinct conformations—one was the normal form, and the other, the disease-causing form. The latter could change the normal prion into an infectious one.

Prusiner and colleagues also described hereditary forms of CJD caused by mutations of prion genes. By introducing mutant genes into the germ lines of transgenic mice, they showed that scrapie-like disease in the mice can be induced. Further, they demonstrated that the 'prion knock-out' transgenic mice were completely resistant to scrapies.

The above findings indicated that as infectious agents, prions were unique. Besides CJD, kuru and scrapies, a long list of conditions is now regarded as caused by prions. They include bovine spongyform encephalopathy (or 'mad cow disease'), a hereditary dementia called Gertsmann-Sträussler-Scheinker, and fatal familial insomnia.

Although prions are now widely regarded as infectious particles capable of causing several disorders, many scientists still believe that the final story is yet to be told, for there may be viruses, hitherto unidentified, that may be responsible for such illnesses. Yet, scientists are unanimous in their approval of Prusiner's enormous contributions to the field of degenerative neurological diseases.

References

1. Aguzzi A, Heppner FL. Pathogenesis of prion diseases: a progress report. Cell Death Differ. 2000 Oct;7(10):889-902.
2. Brown P. The pathogenesis of transmissible spongiform

encephalopathy: routes to the brain and the erection of therapeutic barricades. Cell Mol Life Sci. 2001 Feb;58(2):259-65.

3. Kretzschmar HA, Tings T, Madlung A, Giese A, Herms J. Function of PrP(C) as a copper-binding protein at the synapse. Arch Virol Suppl. 2000;(16):239-49.

4. Masters CL, Beyreuther K. The Worster-Drought syndrome and other syndromes of dementia with spastic paraparesis: the paradox of molecular pathology. J Neuropathol Exp Neurol. 2001 Apr;60(4):317-9.

5. Prusiner, SB. Science 1982, 216:136-44

6. Prusiner SB. Shattuck lecture—neurodegenerative diseases and prions. N Engl J Med. 2001 May 17;344(20):1516-26.

7. Soto C, Saborio GP. Prions: disease propagation and disease therapy by conformational transmission. Trends Mol Med. 2001 Mar;7(3):109-14.

1998

☆

Physics
ROBERT B. LAUGHLIN, HORST L. STORMER and DANIEL C. TSUI
for their discovery of a new form of quantum fluid with fractionally charged excitations.

☆

Chemistry
WALTER KOHN
for his development of the density-functional theory, and
JOHN A. POPLE
for his development of computational methods in quantum chemistry.

☆

Physiology or Medicine
ROBERT F. FURCHGOTT, LOUIS J. IGNARRO and FERID MURAD
for their discoveries concerning nitric oxide as a signaling molecule in the cardiovascular system.

☆

Literature
JOSE SARAMAGO
who with parables sustained by imagination, compassion and irony continually enables us once again to apprehend an elusory reality.

☆

Peace
JOHN HUME and DAVID TRIMBLE
for their efforts to find a peaceful solution to the conflict in Northern Ireland.

☆

Economics
AMARTYA SEN
for his contributions to welfare economics.

1998

ROBERT FRANCIS FURCHGOTT (1911-2008), LOUIS JOSEPH IGNARRO (BORN 1941), AND FERID MURAD (2936-2023)

Robert Francis Furchgott

Louis Joseph Ignarro

Ferid Murad

© The Nobel Foundation

Robert Francis Furchgott (left)**, Louis Ignarro** (middle)**,** and **Ferid Murad** shared the 1998 Nobel Prize in Physiology or Medicine "for the discovery concerning nitric oxide as the signaling molecule in the cardiovascular system."

Robert Furchgott was born in Charleston, South Carolina and earned his Ph.D., from Chicago's Northwestern University in 1940. For much of his career, he has been at State University of New York as professor of pharmacology.

Louis Ignarro was born in Brooklyn, New York. After earning his Ph.D., in pharmacology from the University of Minnesota in 1966, he served at various places, including the University of California in Los Angeles. Ferid Murad was born in Whiting, Indiana. He studied medicine at Western Reserve University School of Medicine in Cleveland, Ohio. He has held positions both in academic and pharmaceutical industries.

From early in his research career, Furchgott studied vascular smooth muscle responses to neurotransmitters and drugs. In 1953, he reported that acetylcholine, a potent in-vivo vasodilator, caused a paradoxical constriction of strips of rabbit aorta. Later he described the light-induced relaxation of arteries (photo-relaxation) and made several important contributions in vascular pharmacology. A chance occurrence in May 1978 led to a breakthrough. In one of his experiments, acetylcholine produced relaxation of aortic smooth muscles. Seeking an explanation for this unexpected finding, Furchgott learned that a technician had "carelessly" prepared an aortic strip, in which the endothelium had been intact. Furchgott then pursued the study of the role of endothelium in vasodilation.

From his famous "sandwich" experiments using aortic strips with and without endothelium, he discovered that an intact endothelium was needed for a vasodilator response to acetylcholine and to other similar drugs. In 1980, Furchgott proposed that an "endothelial dependent relaxing factor" (EDRF) was responsible for such an action.

Ignarro, in collaboration with Furchgott (and independently), attempted to identify EDRF. In a brilliant series of experiments these scientists showed that when hemoglobin solution was exposed to endothelial cells that were stimulated to produce EDRF, the resulting spectral change was like that caused by nitric oxide (NO). In 1986, they suggested that EDRF was a chemical molecule, probably NO, or a similar substance.

Ferid Murad investigated whether the vasodilator action of nitroglycerin was mediated via NO. Bubbling NO through tissues containing guanylyl cyclase (GC) led to an increase in cyclic GMP; Murad surmised that NO stimulated GC to synthesize cyclic GMP, which caused relaxation of myosin in the vascular smooth muscles.

The discovery of NO as a new signaling molecule had important physiological and therapeutic implications. Many extra-vascular effects of NO, such as neuronal signal processing and its role in inflammation and cellular defense, have since been discovered. Inhaled NO and NO-producing drugs are being used in the treatment of persistent pulmonary hypertension in newborn infants and erectile dysfunction in adults.

References

1. Furchgott RF. A Research Trail over Half a Century. Annual Review of Pharmacology and Toxicology. 1995; 35: 1-27

2. Ignarro LJ, Byrns RE, Woods KS. Biochemical and Pharmacological Properties of Endothelium-dependent Relaxing Factor and its Similarity to Nitric Oxide Radical. in Vanhoutte PM (ed): Vasodilation, Vascular Smooth Muscles, Peptides, Autonomic Nerves, and Endothelium. New York, Raven Press. 1998; 427-436

3. Moncada S. The First Robert Furchgott Lecture: From Endothelium-Dependent to L-Arginine: NO Pathway. Blood Vessels 1990; 27:208-217.

4. Zhang J and Snyder SH. Nitric Oxide in the Nervous System. Annual Review of Pharmacology and Toxicology. 1995; 35: 213-33.

1999

☆

Physics
GERADUS'T HOOFT and MARTINUS J. G. VELTMAN
for elucidating the quantum fluid with fractionally charged excitations.

☆

Chemistry
AHMED H ZEWALL
*for his studies of the transition states of chemical
reactions using femtosecond spectroscopy.*

☆

Physiology or Medicine
GÜNTER BLOBEL
*for the discovery that proteins have intrinsic signals that
govern their transport and localization in the cell.*

☆

Literature
GÜNTER GRASS
whose frolicksome fables portray the forgotten face of history.

☆

Peace
MÉDICINS SANS FRONTIÉRES (Doctors Without Borders)

☆

Economics
ROBERT A MUNDELL
*for his analysis of monitory and fiscal policy under different exchange
rate regimes and his analysis of optimum currency areas.*

1999

GÜNTER BLOBEL (1936-2018)

© The Nobel Foundation

Günter Blobel was awarded the 1999 Nobel Prize in Physiology or Medicine "for the discovery that proteins have intrinsic signals that govern their transport and localization in the cell."

Blobel was born in a small farming town called Waltersdorf, in Silesia, Germany. Toward the end of the Second World War, as his family fled to escape the advancing Russian army, they passed by Dresden, and the young Günter Blobel saw for the first time, a large city being ravaged by war.

He earned a medical degree from the University of Tübingen and a doctoral degree in oncology from the University of Wisconsin in Madison. In 1967, he joined New York's Rockefeller University as post-doctoral fellow under George Palade, the 1974 Nobel laureate, and studied cell biology. Blobel became professor at Rockefeller University's Howard Hughes Medical Institute in 1973.

Palade had shown that various proteins synthesized in the common ribosomal pool find themselves at specific destinations in the cell. Blobel began studying the mechanisms by which intracellular proteins manage to find their target locations.

In 1971, Blobel and his colleague David Sabatini proposed the "signal hypothesis," according to which, proteins synthesized in the cell find their targets based upon specific signal recognition sequences. With little experimental proof at the time, the pair made a set of bold and brilliant propositions: 1) signal sequences for protein targeting are amino-terminal portions of proteins; 2) 'signal recognition proteins' (SRP) located on the membrane of endoplasmic reticulum recognize the signal sequences that are attached to cytosolic cofactors; and 3) specific protein-conducting channels (PCCs) will open to allow protein passage across the membranes.

The molecular mechanism of intracellular protein trafficking proposed and later confirmed by Blobel has been compared to "zip codes" and "address tags" used by the post office to route the outgoing mail. Although several refinements have been made to the hypothesis of signal sequence-mediated protein translocation, Blobel's original hypothesis has been proved impeccably accurate. This concept has also helped in understanding molecular mechanisms of such varied conditions as Alzheimer's disease, cystic fibrosis, primary hyperoxyluria and primary hypercholesterolemia. Current and future applications of this concept include commercial manufacturing of drugs such as insulin, growth hormone, erythropoietin and interferon.

"This is an award for the entire field of cell biology" remarked Blobel after hearing of his award. He added that even in this era of "prime-time PCR and cloned sheep," it is cell biology that is entering a golden age. Blobel himself has richly contributed to spearhead such a golden age—generations of scientists he has trained are destined to become leading cell biologists.

While fleeing Germany in 1945, the eight-year-old Blobel had witnessed with horror the bombing and destruction of Dresden's magnificent Baroque-era Church *Frauenkirche*. In 1999, Blobel donated his $960,000 Nobel Prize money to *The Friends of Dresden*, an organization he had founded in 1995 to help rebuild *Frauenkirche* to its original glory.

References

1. Alberts, B. et al (Eds). The Molecular Biology of the Cell. 3rd edition. Garland Publishing, New York 1994.
2. Anderson D, Walter P. Blobel's Nobel: a vision validated. Cell. 1999 Dec 10;99(6):557-8.
3. Birmingham K. A Nobel for Blobel: Nature Medicine. 1999 Nov;5(11):1230.

2000

☆

Physics
ZHORES I ALFEROV
for basic work on information and communication technology developing semiconductor
HERBERT KROEMER
for developing semiconductor heterostructure used in high-speed and opto-electronics; and
JACK S KILBY
for his part in the invention of integrated circuits.

☆

Chemistry
ALAN J HEEGER, ALAN G MacDIARMID and HIDEKI SHIRAKAWA
for the discovery and development of conductive polymers

☆

Physiology or Medicine
ARVID CARLSSON, PAUL GREENGARD and ERIC R KANDEL
for the discovery of signal transduction in the nervous system

☆

Literature
GAO XINGJIAN
for an œuvre of universal validity, bitter insights and linguistic ingenuity, which has opened new paths for the Chinese novel and drama

☆

Peace
KIM DAE JUNG
President of South Korea

☆

Economics
JAMES J HECKMAN
for his development of theory and methods for analyzing selective samples, **and**
DANIEL L McFADEEN
for his theory and methods for analyzing discrete choices.

2000

ARVID CARLSSON (1923-2018),
PAUL GREENGARD (1925-2018),
AND ERIC KANDEL (BORN 1929)

Arvid Carlsson   **Paul Greengard**   **Eric Kandel**

© The Nobel Foundation

Arvid Carlsson (left), **Paul Greengard** (middle), **and Eric Kandel** were awarded the Nobel Prize in Physiology or Medicine for their discoveries concerning "signal transduction in the nervous system"

Carlsson was born in Uppsala, Sweden, and obtained his medical degree in 1951 from the University of Lund, where he worked until 1959. He then moved to Göteborg University as professor of Pharmacology. At Lund, he developed exquisite methods for measuring minute concentrations of dopamine and discovered it to be a new neurotransmitter. He showed that dopamine was found in areas of the brain not containing noradrenaline, hence it was not a precursor of the latter. Dopamine concentrations were also particularly high in the basal ganglia, the part of the brain crucial for the control of motor behavior.

Carlsson made a breakthrough when he showed that injections of reserpine, a natural alkaloid that depletes the stored synaptic transmitters, led to a loss of spontaneous movements in experimental animals. These functions could be restored, however, by treating the animals with L-dopa—a precursor of dopamine which also normalized the tissue concentrations of dopamine. These findings led to the discovery that reduced levels of dopamine in the brain was the basis for symptoms of Parkinson's disease. Soon L-dopa was developed as an efficient remedy for this disease. Carlsson later showed that many drugs used for schizophrenia affected synaptic transmission by blocking dopamine receptors. These works led to additional advances in developing selective serotonin-uptake blockers—a new generation of anti-depressive drugs widely in use today.

Born in New York, Paul Greengard earned his MD degree from Johns Hopkins University in 1953. After postdoctoral studies in biochemistry from the University of London he served at Yale University in New Haven, Connecticut, and at Rockefeller University in New York.

While at Yale, Greengard focused his research on cellular biochemistry of the brain, especially on the phenomenon called "slow synaptic transmission," in which the effect on the neurons would last from minutes to hours, as opposed to seconds. It was known that slow synaptic transmission was responsible for keeping alertness and mood, as well as for modulating such faster functions as speech, movements, and sensory perception. Greengard showed that slow synaptic transmission involved a chemical reaction called protein-phosphorylation in which the phosphate groups are coupled to a protein altering the form and function of the protein.

He found that when dopamine stimulated the receptors in the cell membrane, it increased the cellular concentrations of cyclic AMP, activating the enzyme protein kinase A, which in-turn enabled protein-phosphorylation. This was the basis for the heightened excitability of cell membrane. Thus, Greengard's discoveries on protein phosphorylation helped in understanding the basic mechanism of action of several drugs specifically affecting phosphorylation of proteins in nerve cells.

Eric Kandel was born in Vienna, Austria, and later became a

naturalized citizen of USA. After earning his MD degree from New York University in 1956, he studied and practiced psychiatry at Harvard University before moving to Columbia University, where he became the Director of the Center for Neurobiology and Behavior.

Realizing that the mammalian brain was too complex for the study of memory processing, Kandel chose the sea slug *Aplysia*. This animal has a 'simple brain' with some 20,000 relatively large neurons. Kandel showed that certain stimuli amplified the protective reflex in the sea slug. The amplification also remained for weeks suggesting the phenomenon reflected learning from the stimuli.

He then discovered that learning was essentially a process of amplification of the synapses connecting sensory nerve cells to muscle neurons that were responsible for the protective reflex of the sea slug. Weaker stimuli led to short-term memory lasting for minutes-to-hours. This was secondary to altered ion channels, which allowed excessive calcium influx into the nerve terminal, and caused an amplification of the reflex.

Powerful and long-lasting stimuli, on the other hand, caused long-term memory that remained for weeks. Such stimuli enhanced cellular concentrations of cAMP and protein kinase A. The result was that the shape of the synapse had changed, thereby creating a long-lasting increase of synaptic functions. In contrast to short-term memory, long-term memory, required that new proteins are formed. By preventing the synthesis of new proteins, one could block long-term memory—but not the short-term memory.

Thus, Kandel showed that both short and long-term memory functions were "located in the synapses" and changes in the synaptic functions were central for the formation of different types of memories. While we are far from understanding the nature of memory function, Kandel's work was an important foundation for future research in this field. Many laboratories are pursuing work on the physiology of memory functions as well on developing treatments for memory dysfunctions.

References

1. De Camilli P, Carew TJ. Nobel Celebrates the Neurosciences: Modulatory signaling in the Brain. Cell. 103:829-833; 2000.

2001

Physics
Eric A. Cornell, Wolfgang Ketterle and Carl E. Wieman

for the achievement of Bose-Einstein condensation in dilute gases of alkali atoms, and for early fundamental studies of the properties of the condensates.

Chemistry
William S. Knowles, Ryoji Noyori, and K. Barry Sharpless

for their work on chirally catalysed hydrogenation reactions (to Knowles and Noyori) for his work on chirally catalysed oxidation reactions (to Sharpless).

Physiology or Medicine
Leland H. Hartwell, Tim Hunt and Sir Paul M. Nurse

for their discoveries of key regulators of the cell cycle.

Literature
Sir Vidiadhar Surajprasad Naipaul

for having united perceptive narrative and incorruptible scrutiny in works that compel us to see the presence of suppressed histories.

Peace
The United Nations (UN) and Kofi Annan

for their work for a better organized and more peaceful world.

Economics
George A. Akerlof, A. Michael Spence and Joseph E. Stiglitz

for their analyses of markets with asymmetric information.

2001

LELAND HARRISON HARTWELL (BORN 1939), SIR RICHARD TIMOTHY HUNT (BORN 1943) AND SIR PAUL MAXIME NURSE (BORN 1949)

Leland Harrison Hartwell

Sir Richard Timothy Hunt

Sir Paul Maxime Nurse

Leland H. Hartwell, **Tim Hunt**, and **Sir Paul M. Nurse** received the Nobel Prize for Physiology or Medicine for their discoveries of key regulators of the cell cycle.

A human adult has approximately 100,000 billion cells, and these are not static. Each cell continues to grow, then divides, producing a duplicate daughter cell before dying. These processes are called cell cycles, and they are highly coordinated.

Leland Hartwell discovered a specific class of genes that control the cell cycle. One such gene, "start," has a central role in controlling the first step of each cell cycle. Hartwell also introduced the concept of "checkpoint," a valuable aid to understanding the cell cycle.

Paul Nurse used genetic and molecular methods to identify, clone, and characterize cyclin-dependent kinase (CDK), one of the key regulators of the cell cycle. He showed that CDK function is highly conserved during evolution. CDK drives the cell through the cell cycle by chemical modification, a process called phosphorylation of other proteins. Timothy Hunt discovered cyclins, which are proteins that regulate the CDK function. He showed that cyclins are degraded periodically at each cell division, a mechanism that proved to be of general importance for cell cycle control.

Hartwell notes that by the time he was 10 or so, he was curious about the natural world. An avid collector of bugs, butterflies, lizards, snakes, and spiders, a painful experience reinforced his conviction not to believe everything he reads. It seems that he had read somewhere that lizards had no teeth. To test it, he opened the mouth of a big lizard he had caught. The lizard not only bared its fine set of teeth but sank them into his thumb!

Harwell says that during his adolescence, some outstanding teachers steered his interest in science. He entered the California Institute of Technology and worked in several prestigious schools. He says that he was greatly influenced by the works of Francois Jacob, Jacques Monod, and Renato Dulbecco, all Nobel laureates, stimulating him to continue research in cell biology.

Tim Hunt grew up in London decades after the Second World War. Like all Britons, he too faced food rationing. Hunt started studying Latin and Greek but found them to be hard. Influenced by good teachers, he developed a love for biology, chemistry, and the sciences. He joined Clare College in Cambridge. After his PhD, he worked for a short while at New York's Albert Einstein College of Medicine before returning to Cambridge, where he continued research in developmental and cell biology, eventually discovering cyclins. He defined a large family of cyclins in *Xenopus* (an aquatic frog native to sub-Saharan Africa) and demonstrated their role in the cell cycle.

Upon hearing his name as a Nobel Prize recipient, Kerstin Westin of Uppsala wrote to Hunt, wondering whether he was the same Hunt in whose family home she served as an au pair girl in Oxford in 1950. She was! Ms. Westin had not predicted a glorious future for the 7-year-old Hunt, but she was delighted to meet him 51 years later in Stockholm.

Paul studied at Lyon Park and received his BSc degree in Biology in 1970 from the University of Birmingham, his PhD degree in 1973 from the University of East Anglia for research on *Candida Utilise*. He then pursued postdoctoral work at the University of Bern, the University of Edinburgh and the University of Sussex. In 1976, Nurse identified the gene called cdc2, which controls the progression of cell cycles in two critical steps.

Paul Nurse's family was from Norfolk. Paul's mother was 18 when she became pregnant. Fearing a label of 'illegitimacy' for the child, her parents sent her away to her aunt's house for delivery. After Paul's birth, his grandmother declared that she was the mother of her "new son." Thus, Paul grew up thinking that his grandmother was his mother, and his biological mother was his sister.

He learned the truth about his mother in 2006, serendipitously, when the US government declined his green card visa application because of an "error" in his mother's name. Surprised, Dr. Nurse obtained his original British birth certificate, where he saw that the person he had thought was his sister was indeed his biological mother. But by then, his parents, grandparents, and other close relatives had died—therefore, he never learned the identity of his father.

In an addendum to his Nobel Lecture added in 2008, Nurse described the above information. In it, he says, "...then there is the final irony that even though I am a geneticist, my family managed to keep my genetic origins secret from me for over half a century."

2002

☆

Physics
Raymond Davis Jr., Masatoshi Koshiba, Riccardo Giacconi

*for pioneering contributions to astrophysics, in particular for
the detection of cosmic neutrinos* (to Davis Jr., and Koshiba)
and *for pioneering contributions to astrophysics, which have led
to the discovery of cosmic X-ray sources* (to Giacconi).

☆

Chemistry
John B. Fenn, Koichi Tanaka, and Kurt Wüthrich

*for the development of methods for identification and
structure analyses of biological macromolecules.*

☆

Physiology or Medicine
Sydney Brenner, H. Robert Horvitz and John E. Sulston

*for their discoveries concerning genetic regulation of
organ development and programmed cell death.*

☆

Literature
Imre Kertész

*for writing that upholds the fragile experience of the
individual against the barbaric arbitrariness of history.*

☆

Peace
Jimmy Carter

*for his decades of untiring effort to find peaceful solutions
to international conflicts, to advance democracy and human
rights, and to promote economic and social development.*

☆

Economics
Daniel Kahneman and Vernon L. Smith

for having integrated insights from psychological research into economic science, especially concerning human judgment and decision-making under uncertainty (to Kahneman) and *for having established laboratory experiments as a tool in empirical economic analysis, especially in the study of alternative market mechanisms* (to Smith).

2002

Sydney Brenner (1927-2019), Howard Robert Horvitz (born 1947) and John E. Sulston (1942-2018)

Sydney Brenner Howard   **Robert Horvitz**   **Sir John Edward Sulston**

© The Nobel Foundation

Sydney Brenner, **H. Robert Horvitz**, and **John E. Sulston** were awarded the Nobel Prize in Physiology or Medicine for their discoveries concerning the genetic regulation of organ development and programmed cell death.

The human body consists of hundreds of cell types that mature and differentiate into specialized cells in body tissues and organs. Over time, these cells generate new cells, and the old cells die—a process called programmed cell death, which is the culmination of a meticulously articulated series of steps. The 2002 Nobel Prize winners in Physiology or Medicine made seminal discoveries concerning the genetic regulation of organ development and programmed cell death.

In the early 1960s, Sydney Brenner realized cell differentiation and organ development were hard to tackle in higher animals. Therefore, he considered the roundworm *Caenorhabditis elegans* for his research. This approximately 1 mm-long worm has a short generation time and is transparent, which makes it possible to follow cell division directly under the microscope. By 1974, he had shown specific gene mutations that could be induced in the genome of *C. elegans* by the chemical compound EMS (ethyl methane sulphonate).

John Sulston extended Brenner's work with *C. elegans* and developed techniques to study cell divisions in the nematode, from the fertilized egg to the 959 cells in the adult organism. In a 1976 publication, Sulston described the cell lineage for a part of the developing nervous system. He showed that the cell lineage is invariant, i.e., every nematode underwent the same program of cell division and differentiation.

As a result of these findings, Sulston made the seminal discovery that specific cells in the cell lineage always die through programmed cell death, and that one can monitor this in living organisms. He described the visible steps in the cellular death process and demonstrated the first mutations of genes participating in programmed cell death, including the *nuc-1* gene. Sulston also showed that the protein encoded by the *nuc-1* gene degrades the DNA of the dead cell.

Robert Horvitz continued Brenner's and Sulston's work on the genetics and cell lineage of *C. elegans*. In a series of elegant experiments that started during the 1970s, Horvitz used *C. elegans* to investigate whether there was a genetic program controlling cell death. In a paper published in 1986, he identified the first two bona fide "death genes," *ced-3* and *ced-4*. He showed that functional *ced-3* and *ced-4* genes were a prerequisite for cell death to be executed.

Later, Horvitz showed that another gene, *ced-9*, protects against cell death by interacting with *ced-4* and *ced-3*. He also identified several genes that direct how the dead cell is eliminated. Horvitz showed that the human genome contains a *ced-3*-like gene. We now know that most genes involved in controlling cell death in *C. elegans* have counterparts in humans.

Brenner was born in Germiston in the Gauteng province in South Africa. At age 15, he joined the University of Witwatersrand to study medicine, but the authorities noted that he would be too young to practice medicine upon graduation. Thus, he was required to complete a BSc degree in anatomy and physiology. He then rejoined the medical school, where his performance was poor. He failed in medicine and nearly failed in surgery. However, he obtained his medical degree in 1951. During his BSc training, Brenner developed a deep liking for anatomy and physiology rather than clinical practice. With the help of a scholarship, he studied at the University of Oxford and completed a DPhil degree.

Brenner was a brilliant writer. From 1994 to 2000, he wrote a column "Loose Ends," for the *Current Biology* journal. He renamed it "False Starts" in 1998, when the journal moved the column from the back of the journal to the front. A collection of those essays published by Current Biology LTD (and a paperback by BioMed Central) is a classic. Brenner organized a seminar series in Singapore where many leading scientists gave lectures on topics related to biology, evolution, and the history of biological sciences. Those lectures including the Brenner's talks were adapted in a volume titled *In the Spirit of Science: Lectures by Sydney Brenner* in 2017.

Brenner made several fundamental contributions to the field of molecular biology, including the discovery of transfer RNA (tRNA) and the concept of messenger RNA.

2003

Physics
Alexei A. Abrikosov, Vitaly L. Ginzburg and Anthony J. Leggett
for pioneering contributions to the theory of superconductors and superfluids.

Chemistry
Peter Agre and Roderick MacKinnon
for the discoveries concerning channels in cell membranes (to Agre)
and *for the discovery of water channels and with one half to for
structural and mechanistic studies of ion channels* (to MacKinnon).

Physiology or Medicine
Paul C. Lauterbur and Sir Peter Mansfield
for their discoveries concerning magnetic resonance imaging.

Literature
John M. Coetzee
who in innumerable guises portrays the surprising involvement of the outsider.

Peace
Shirin Ebadi
*for her efforts for democracy and human rights. She has focused
especially on the struggle for the rights of women and children.*

Economics
Robert F. Engle III and Clive W.J. Granger
*for methods of analyzing economic time series with time-
varying volatility (ARCH) and for methods of analyzing
economic time series with common trends (cointegration).*

2003

PAUL CHRISTIAN LAUTERBUR (1929-2007) AND PETER MANSFIELD (1933-2017)

Paul Christian Lauterbur　　**Sir Peter Mansfield**

Paul C Lauterbur and **Peter Mansfield** were awarded the Nobel Prize in Physiology or Medicine for their discoveries concerning magnetic resonance imaging.

Atomic nuclei in a strong magnetic field rotates with a frequency dependent on the magnetic field's strength. Their energy can be increased if they absorb radio waves with the same frequency (resonance). Radio waves are emitted when the atomic nuclei return to their previous energy level. These discoveries were awarded the Nobel Prize in Physics in 1952. However, it was Lauterbur and Mansfield's pioneering contributions that led to the applications of magnetic resonance in medical imaging.

Paul Lauterbur discovered the possibility of creating a two-dimensional picture by introducing gradients in the magnetic field. He could determine the origin of the emitted radio waves by analyzing their characteristics. This approach made it possible to build up two-dimensional pictures of structures that he could not visualize with other methods.

Peter Mansfield further developed the utilization of gradients in the magnetic field. He showed how the signals could be mathematically analyzed, which made creating a useful imaging technique possible. Mansfield also showed how extremely fast imaging could be achieved. This became technically possible later within a decade.

Lauterbur was born and raised in Sidney, Ohio. While in high school, he built his own laboratory in the basement of his parents' house. His chemistry teacher in school encouraged his experimentation and allowed him to do his experiments at the back of the class. In 1950, he was drafted into the United States Army, where he continued his nuclear magnetic resonance machine research.

After obtaining his BS in chemistry from Case Western Reserve University in Ohio, Lauterbur graduated and received his PhD from the University of Pittsburgh. He then worked at Stony Book University and Stanford University before moving to the University of Illinois in 1985.

Lauterbur credits the idea of the MRI to a brainstorm one day at a suburban eatery, where he scribbled the first model on a table napkin. He followed his brainstorm by conducting experiments, which led to the work that would earn him a Nobel Prize.

Although nuclear magnetic resonance (NMR), the scientific principle underlying MRI, had been developed during the final decades of World War II, only after the research of Lauterbur and Mansfield on NMR technology did other scientists adapt it for developing MRI.

Mansfield was born in Lambeth, a south London district in the United Kingdom. During World War II, he was evacuated from London to various places but returned to London after the war. His school asked him to take the "11 +" examination, which he had never heard of. With no training, Mansfield failed the examination and could not enter the local grammar school. But his marks were high enough for him to get an entry to a school in Peckham. But at age 15, one of his teachers told him he was not good enough for science. Mansfield left school and began working as a printer's assistant.

At 18, having developed an interest in rocketry, Mansfield took up a job with the Rocket Propulsion Department of the Ministry of Supply in Westcott, Buckinghamshire. He was then recruited to serve in the army for two years. Upon returning from the military, Mansfield continued his studies by attending night schools. He then entered Queen Mary College, University of London, where he studied physics and obtained his BSc in 1959 and his PhD in 1962. His thesis was titled *Proton magnetic resonance relaxation in solids by transient methods*.

Following his PhD, Mansfield completed his postdoctoral research at the University of Illinois at Urbana-Champaign campus before returning to England in 1964.

Imaging human internal organs is very important for medical diagnosis, treatment, and follow-up. Thus, Lauterbur and Mansfield's fundamental research led to the development of modern MRI techniques, which have made a huge impact on all branches of medical science.

2004

☆

Physics
David J. Gross, H. David Politzer and Frank Wilczek
for the discovery of asymptotic freedom in the theory of the strong interaction.

☆

Chemistry
Aaron Ciechanover, Avram Hershko and Irwin Rose
for the discovery of ubiquitin-mediated protein degradation.

☆

Physiology or Medicine
Richard Axel and Linda B. Buck
for their discoveries of odorant receptors and the organization of the olfactory system.

☆

Literature
Elfriede Jelinek
for her musical flow of voices and counter-voices in novels and plays that with extraordinary linguistic zeal reveal the absurdity of society's clichés and their subjugating power.

☆

Peace
Wangari Muta Maathai
for her contribution to sustainable development, democracy and peace.

☆

Economics
Finn E. Kydland and Edward C. Prescott
for their contributions to dynamic macroeconomics: the time consistency of economic policy and the driving forces behind business cycles.

2004

RICHARD AXEL (BORN 1946) AND
LINDA BROWN BUCK (BORN 1947)

Richard Axel   **Linda Brown Buck**

Richard Axel and **Linda Buck** were awarded the Nobel Prize in Physiology or Medicine for their discoveries of odorant receptors and the organization of the olfactory system.

How do animals recognize and remember odors? This has been a mystery for a long time. Axel and Buck's pioneering studies clarified how the human olfactory system works.

They discovered a large family of genes that give rise to olfactory receptors in the upper part of the nasal epithelial surface. These receptors can detect inhaled odorant molecules. Moreover, each receptor recognizes only one type of odor—thus, they are highly specialized. From the surface of the nasal epithelium, messages are passed along highly specialized, thin nerve fibers that travel to the brain region called the olfactory bulb—the primary brain region where the sense of smell is first received.

The odor information then gets passed along to at least three important brain regions—the amygdala, orbitofrontal cortex, and the hippocampus. These stations then interpret the odor, eliciting emotional responses, triggering memory, and stimulating learning. Examples of odors eliciting emotional responses are many, but to cite a couple: the smell of a rose in a flower shop might trigger the memory of a Valentine's Day gift from a friend. The earthy smell after the rain on a hot summer day (known as petrichor) may evoke nostalgic memories of happy days and coziness. Richard Axel and Linda Buck published their initial paper in 1991 describing a large family of about one thousand genes for odorant receptors. Then, working independently, they carried out elegant studies that clarified the processes of olfactory systems, from the molecular to higher macro levels.

Odor and taste are also closely connected. Tasting a good wine can trigger an array of odorant experiences. Odors can also trigger negative emotions. Tasting a spoiled fruit or vegetable during childhood could lead to a permanent dislike of the odors from that fruit or vegetable. Conditions where the sense of smell is lost lead to serious handicaps.

Detection of smell has a survival advantage. Animals need to identify edible food by its smell and avoid those that can cause harm or death. Alex and Buck studied the smell functions in mice, which have about one thousand odor receptors. Humans have fewer such receptors than mice, probably because of their loss during evolution. They also showed that the olfactory epithelium of dogs is nearly 40 times larger, which explains the exquisite olfactory abilities of dogs.

Newborn mammals need smell to find their mother's teats to obtain milk. Piglets are essentially blind at birth; therefore, they need their sense of smell to survive. When their smell sensations are experimentally blocked, piglets wander around their mother's udder, fail to find their mother's teat, and may die from hunger.

Axel was born in New York City to Polish Jewish immigrants. He received his BA degree from Columbia University and his MD degree from Johns Hopkins University. Although he performed poorly in clinical medicine, his professors urged him to pursue research in clinical medicine. He returned to Columbia University and became a full professor in 1978.

Linda Buck was born in Seattle, Washington. She received her BS degree in psychology and microbiology from the University of Washington in 1975 and earned her PhD in immunology in 1980. She joined Richard Axel's laboratory during her postdoctoral research at Columbia University. After reading a research paper by Sol Snyder and his group at Johns Hopkins University, Buck pursued her work to map olfactory processes at the molecular level. Buck and Axel worked with rat genes, identified a family of genes that code for more than 1000 odor receptors, and published these findings in 1991.

The general principles Axel and Buck discovered for the olfactory system also apply to other sensory systems. Pheromones are molecules that can influence different social behaviors, especially in mammals. Axel and Buck discovered that pheromones are detected by two other families of G protein-coupled receptors localized to a different part of the nasal epithelium. The tongue's taste buds have yet another family of G protein-coupled receptors associated with the sense of taste. These discoveries show how sense of smell and taste are connected and how they operate in the brain, triggering memory and influencing behavior.

2005

☆

Physics
Roy J. Glauber, John L. Hall and Theodor W. Hänsch
for his contribution to the quantum theory of optical coherence
(to Glauber); and *for their contributions to the development
of laser-based precision spectroscopy, including the optical
frequency comb technique* (to Hall and Hänsch).

☆

Chemistry
Yves Chauvin, Robert H. Grubbs and Richard R. Schrock
for the development of the metathesis method in organic synthesis.

☆

Physiology or Medicine
Barry J. Marshall and J. Robin Warren
*for their discovery of the bacterium Helicobacter pylori and
its role in gastritis and peptic ulcer disease.*

☆

Literature
Harold Pinter
*who in his plays uncovers the precipice under everyday
prattle and forces entry into oppression's closed rooms.*

☆

Peace
International Atomic Energy Agency (IAEA) and Mohamed ElBaradei
*for their efforts to prevent nuclear energy from being used
for military purposes and to ensure that nuclear energy for
peaceful purposes is used in the safest possible way.*

☆

Economics
Robert J. Aumann and Thomas C. Schelling
*for having enhanced our understanding of conflict and
cooperation through game-theory analysis.*

2005

Barry James Marshall (born 1951) and John Robin Warren (1937-1924)

Barry James Marshall   **John Robin Warren**

Barry J. Marshall and **J. Robin Warren** were jointly awarded the Nobel Prize in Physiology or Medicine for their discovery of *Helicobacter pylori* and their role in gastritis and peptic ulcer disease.

Robin Warren was a pathologist from Perth, Australia. He observed tiny, curved organisms colonizing the lower part of the stomach in about 50% of patients suffering gastritis from whom biopsies had been taken. He made the crucial observation that signs of inflammation were always present in the stomach surfaces closest to where he saw the bacteria under the microscope.

Barry Marshall, a clinical fellow, got interested in Warren's findings, and together, they initiated a study of biopsies from 100 patients. After several attempts, Marshall cultivated a hitherto unknown bacterial species, later named *Helicobacter pylori,* from several of these biopsies.

Together, they found that the organism was present in almost all patients with gastric inflammation, duodenal ulcer, or gastric ulcer. Based on these results, they proposed that *Helicobacter pylori* are involved in causing these diseases.

For a long time, peptic ulcer disorders were considered to result from stress and lifestyle coupled with eating spicy food. The discoveries by Marshall and Warren firmly established that over 90%of peptic ulcers and up to 80% of gastric ulcers were caused by *Helicobacter pylori*.

Marshall was born in Kalgoorlie, Western Australia, but his family moved to Perth when he was eight years old. After studying at Marist College and completing his medical education at the University of Western Australia School of Medicine, he earned his basic medical degree, MBBS, in 1974. While working as a Registrar of Medicine at Royal Perth Hospital, he met Dr Warren, who was studying gastritis. Together, they studied the presence of a spiral bacteria associated with gastritis.

John Warren was born in North Adelaide and obtained his MBBS medical degree from the University of Adelaide. After serving at the Royal Adelaide Hospital and Royal Melbourne Hospital, he joined as a senior pathologist at the Royal Perth Hospital, where he pursued his research studies on gastric and peptic ulcers.

In 1982, Marshall and Warren performed the initial culture of *H. pylori* and developed their hypothesis on the bacterial cause of peptic ulcers and gastric cancer. Their discovery was dismissed initially because, until then, scientists believed that the acidity in the stomach prevented bacterial colonization. With the help of research funding for one year, Marshall and Warren attempted to culturing the bacterium from biopsies obtained from patients having peptic ulcers or gastric ulcers.

A serendipitous event led to their success in culturing the elusive *H. pylori*. As a standard practice, the lab technicians would discard the biopsy specimen after two days of no bacterial growth, because of the assumption that any bacteria that grow after two days is due to contamination. Thus, the lab technicians threw away 30 initial bacterial culture plates after two days of negative cultures. On a week with heavy workload, the technicians forgot to throw away a single culture plate on a Thursday, which stayed until the following Monday. In that sample, Warren and Marshal discovered the hitherto unseen bacteria *H. pylori*. They later learned that *H. pylori* grow more slowly than other microbes.

In 1983, they submitted their findings to the Gastroenterological Society of Australia. However, the reviewers turned down their paper, rating it in the bottom 10% of those received that year.

To prove that *H. pylori* indeed caused gastritis, Marshall undertook a daring self-experimentation. He had a baseline endoscopy of his stomach to prove it was healthy. Then, he drank a broth containing cultured *H. pylori*. Three days later, he developed nausea and halitosis— bad breath—due to a drop in stomach acid production. On days five through eight, he developed vomiting. On the eighth day, he underwent a repeat endoscopy, which showed massive inflammation of his stomach. A culture of his stomach biopsy taken at that time showed *H. pylori*, proving that the bacteria had colonized his stomach. On the 14th day, he underwent a third endoscopy and began a course of antibiotics, which cured his illness.

Marshall's illness and recovery were classic examples proving "Koch's Postulates." proposed by the German Scientists Robert Koch and Frederick Loeffler in 1884. Four requirements were needed to establish a causal relationship between a microbe and a disease. They were: 1) the pathogen must be found in all organisms suffering from the disease but not in healthy organisms; 2) The pathogen must be isolated from a diseased organism and grown in pure culture; 3) the cultured pathogen should cause the disease when introduced into a healthy organism; 4) the pathogen must be re-isolated from the inoculated, diseased new experimental host and identified as identical to the original pathogen causing the disease.

Marshall continued his research at the Royal Perth Hospital and the University of Virginia, after which he worked at the University of Western Australia.

Now, doctors routinely prescribe a course of antibiotics that cures gastric and peptic ulcers, preventing unnecessary acid-reducing medications and surgical procedures that include resections of large parts of the stomach and duodenum.

2006

☆

Physics
John C. Mather and George F. Smoot
*for their discovery of the blackbody form and anisotropy
of the cosmic microwave background radiation.*

☆

Chemistry
Roger D. Kornberg
for his studies of the molecular basis of eukaryotic transcription.

☆

Physiology or Medicine
Andrew Z. Fire and Craig C. Mello
*for their discovery of RNA interference - gene
silencing by double-stranded RNA.*

☆

Literature
Orhan Pamuk
*who in the quest for the melancholic soul of his native city has
discovered new symbols for the clash and interlacing of cultures.*

☆

Peace
Muhammad Yunus and Grameen Bank
for their efforts to create economic and social development from below.

☆

Economics
Edmund S. Phelps
for his analysis of intertemporal tradeoffs in macroeconomic policy.

2006

ANDREW ZACHARY FIRE (BORN 1959) AND CRAIG CAMERON MELLO (BORN 1960)

Andrew Zachary Fire

Craig Cameron Mello

Andrew Z. Fire and **Craig C. Mello** were awarded the Nobel Prize in Physiology or Medicine for their discovery of RNA interference, which is gene silencing by double-stranded RNA.

The genes in our body send instructions for the manufacture of proteins from DNA in the cell's nucleus to the protein-synthesizing machinery in the cytoplasm (outside the nucleus but within the cell). These instructions are conveyed by messenger RNA (mRNA).

In 1998, Andrew Fire and Craig Mello published their discovery of a mechanism that can degrade mRNA from a specific gene. This mechanism, RNA interference, is activated when RNA molecules occur as double-stranded pairs in the cell. This discovery helped us understand how DNA is enabled in protein synthesis.

Andrew Zachary Fire was born in Santa Clara County, California.

He studied at the University of California at Berkeley and received an AB degree in Mathematics in 1978. Fire then entered the PhD program in Biology at the Massachusetts Institute of Technology as a National Science Foundation Fellow in the Fall of 1978. Fire continued his training with a fellowship at the Medical Research Council Laboratory of Molecular Biology in Cambridge, England. During his last year at the MRC lab, Fire applied for and secured a research position at the Carnegie Institution in Baltimore, Maryland, and a research grant from the National Institutes of Health for researching gene regulation during the early development of *C. elegans.* In 1986, he moved to Baltimore. Later, he obtained an adjunct faculty appointment at Johns Hopkins University before moving to Stanford University in 2003.

Craig Cameron Mello was born in New Haven, Connecticut. His father earned a PhD in paleontology, and his mother was an artist. Craig Mello grew up in a family that cultivated a strong tradition of discussions around the dinner table, an enriching experience during childhood and beyond. Despite some difficulties during early school years, Mellow dreamed of becoming a scientist. He attended Brown University, majoring in biochemistry and molecular biology.

After obtaining his degree from Brown, he pursued graduate studies in molecular, cellular, and developmental biology at the University of Colorado, Boulder. However, because his mentor had moved to pursue an industry job, Craig moved to Harvard University, where he obtained his PhD.

Craig Mello recounts the events on the day when he heard the news of having won the Nobel Prize. At 4:30 AM in Boston, the "green light" of his house phone began blinking, and his wife told him not to answer because it is likely to be a prank call. A while later, when he had been away, the phone rang again, and this time, his wife answered. Upon his return, she told him that someone played a bad joke, saying he had won the Nobel Prize. When Craig told his wife that, in fact, that very day, the Nobel Committees were planning to announce the Nobel Prize winners, he says that "her jaw dropped." When the phone rang again, he answered it; the voice on the other end told him to get dressed and that in half an hour, his life was about to change.

RNA interference is seen in plants, animals, and humans. It helps regulate gene expression, participates in defense against viral infections, and keeps jumping genes under control. RNA interference has become an important tool for studying how genes function, a task of fundamental importance in the development of medications for treating human diseases.

2 0 0 7

☆

Physics
Albert Fert and Peter Grünberg
for the discovery of Giant Magnetoresistance.

☆

Chemistry
Gerhard Ertl
for his studies of chemical processes on solid surfaces.

☆

Physiology or Medicine
Mario R. Capecchi, Sir Martin J. Evans and Oliver Smithies
*for their discoveries of principles for introducing specific gene
modifications in mice by the use of embryonic stem cells.*

☆

Literature
Doris Lessing
*that epicist of the female experience, who with skepticism, fire and
visionary power has subjected a divided civilization to scrutiny.*

☆

Peace
Intergovernmental Panel on Climate Change
(IPCC) and Albert Arnold (Al) Gore Jr.
*for their efforts to build up and disseminate greater knowledge
about man-made climate change, and to lay the foundations for
the measures that are needed to counteract such change.*

☆

Economics
Leonid Hurwicz, Eric S. Maskin and Roger B. Myerson
for having laid the foundations of mechanism design theory.

2007

MARIO RAMBERG CAPECCHI (BORN 1937) SIR MARTIN JOHN EVANS (BORN 1941) AND OLIVER SMITHIES (1925-2017)

Mario R. Capecchi   **Sir Martin John Evans**   **Oliver Smithies**

Mario R. Capecchi, Martin J. Evans, and Oliver Smithies were awarded the Nobel Prize in Physiology or Medicine for their groundbreaking discoveries concerning the principles for introducing specific gene modifications into mice using embryonic stem cells. Their discoveries led to *gene targeting* technology, which is now applied in all areas of basic and applied biomedical research.

Our DNA is packaged in chromosomes within the nucleus. Chromosomes occur in pairs, one inherited from the mother and the other from the father. The exchange of DNA sequences within such chromosome pairs increases genetic variation in the population and occurs by a process called *homologous recombination*. This process is conserved throughout evolution and was demonstrated in bacteria by

Joshua Lederberg more than 50 years ago, for which he received the 1958 Nobel Prize in Physiology or Medicine.

Mario Capecchi was born in Verona, Italy. In 1941, he and his mother lived near Bolzano, a city in Northern Italy, about 160 miles from Reggio Emilia, another Northern Italian town where his father lived. His mother was active in anti-Fascist movements, for which the Fascist government arrested her that year. She had anticipated such a possibility; she had planned for a family member to take care of her son. However, during much of his childhood, Mario had to live in an orphanage until the end of the war, suffering from malnutrition. After the war, after searching for him for over a year, his mother found him on his ninth birthday in a hospital in Reggio Emilia. She took him to Rome, and then, with help from a family member, she emigrated to the United States and lived in Pennsylvania.

Mario graduated from a Quaker boarding school in Bucks County, Pennsylvania. In 1956, he earned his BSc from Antioch College, Ohio. He then studied at the Massachusetts Institute of Technology. Because of his newly developed liking for molecular biology, he transferred to Harvard University and worked under James D. Watson, co-discoverer of the molecular structure of nucleic acids for which he shared the 1962 Nobel Prize in Physiology of Medicine. Mario earned his PhD in biophysics from Harvard University. After announcing Capecchi's Nobel Prize, his long-lost half-sister, Ms. Marlene Bonelli, contacted him. The siblings had a happy reunion.

Oliver Smithies was born in Halifax, West Yorkshire, England. In 1951, Smithies earned his master's degree and a PhD in biochemistry from the University of Oxford. He then moved to the United States to continue research at the University of Wisconsin. The following year, he joined the Connaught Medical Research Laboratory at the University of Toronto, where he developed the starch gel electrophoresis technique, an advanced method of separating and identifying blood proteins.

In the 1980s, Smithies focused research on gene therapy to treat hereditary diseases. After learning Evan's work on mice embryonic stem cells, Smithies obtained a sample of stem cells from Evans. In those samples, Smithies demonstrated that targeted removal or alteration of genes allowed for the controlled manipulation of the genome in the mouse. He and Capecchi used that breakthrough to breed mice with specific disease conditions. In 1991, Smithies and his laboratory created a "knockout mouse"—so named because one of its genes had been experimentally replaced or "knocked out"—that accurately modeled cystic fibrosis disease.

Martin John Evans was born in Stroud, Gloucestershire, a county in Southwest England. He hailed from a family of scientists, musicians, poets, and teachers. Early on, Evans developed an interest in science. After schooling, with the help of a major scholarship, he joined Christ's College in Cambridge. There, he attended seminars by Sydney Brenner and lectures by Jacques Monod, both Nobel Laureates. Evans earned his BA in 1963, and moved to University College, London, where he worked as a research assistant before earning a PhD in 1969. In 1978, he moved to the Department of Genetics at the University of Cambridge, where he used blastocysts for isolating embryonic stem cells. His research in mouse embryonic stem cells eventually led him and his collaborators to create transgenic mice to use as experimental models to study human illnesses.

Mario Capecchi and Oliver Smithies both had the vision that homologous recombination could be used to modify genes specifically in mammalian cells, and they worked consistently towards this goal. They also showed that homologous recombination can occur between native DNA and introduced DNA, allowing one to develop a process to repair defective genes.

With gene targeting, almost any type of DNA modification in the mouse genome is now possible, allowing scientists to establish the roles of individual genes in health and disease. Gene targeting has already produced more than five hundred different mouse models of human disorders, including cardiovascular and neurodegenerative diseases, diabetes, and cancer.

2008

☆

Physics
Yoichiro Nambu, Makoto Kobayashi and Toshihide Maskawa
*for the discovery of the mechanism of spontaneous broken symmetry
in subatomic physics,* (to Kobayashi and Maskawa) and *for the
discovery of the origin of the broken symmetry which predicts the
existence of at least three families of quarks in nature* (to Maskawa).

☆

Chemistry
Osamu Shimomura, Martin Chalfie and Roger Y. Tsien
for the discovery and development of the green fluorescent protein, GFP.

☆

Physiology or Medicine
Harald zur Hausen, Françoise Barré-Sinoussi and Luc Montagnier
*for his discovery of human papilloma viruses causing
cervical cancer* (to Hausen) and *for their discovery of human
immunodeficiency virus* (to Barré-Sinoussi and Montagnier).

☆

Literature
Jean-Marie Gustave Le Clézio
*author of new departures, poetic adventure and sensual ecstasy,
explorer of a humanity beyond and below the reigning civilization.*

☆

Peace
Martti Ahtisaari
*for his important efforts, on several continents and over more
than three decades, to resolve international conflicts.*

☆

Economics
Paul Krugman
for his analysis of trade patterns and location of economic activity.

2008

Harald zur Hausen Françoise (1936-2023), Françoise Barré-Sinoussi (born 1947) and Luc Montagnier (1932-2022)

Harald zur Hausen   **Françoise Barré-Sinoussi**   **Luc Montagnier**

One-half of the 2008 Nobel Prize in Physiology or Medicine went to Harald zur Hausen for his discovery of human papillomavirus (HPV) causing cervical cancer, and the other half was divided between Françoise Barré-Sinoussi and Luc Montagnier for their discovery of the human immunodeficiency virus (HIV).

Harald Zur Hausen postulated that oncogenic HPV caused cervical cancer, the second most common cancer among women. He realized that HPV-DNA could exist in a non-productive state in the tumors and should be detectable by specific searches for viral DNA. He found HPV to be a heterogeneous family of viruses. Only some types of HPV cause cancer. His discovery led to the characterization of the natural history of HPV infection, the understanding of mechanisms of HPV-induced carcinogenesis, and the development of prophylactic vaccines against HPV acquisition.

Françoise Barré-Sinoussi and Luc Montagnier discovered HIV. They identified virus production in lymphocytes from patients with enlarged lymph nodes in the early stages of acquired immunodeficiency syndrome (AIDS) and in blood from patients with late stages of the disease. Based on its morphological, biochemical, and immunological properties, they characterized this retrovirus as the first known human lentivirus. HIV impairs the immune system because of massive virus replication and cell damage to lymphocytes. The discovery was a seminal step toward understanding the biology of the disease and its antiretroviral treatment.

Harald zur Hausen was born in Gelsenkirchen-Buer, Germany, during World War II. He and his close family survived the war and post-war periods. Even as a child, Hausen was interested in animals and plants—thus, he had determined to become a scientist. After high school, he enrolled at the University of Bonn to study medicine, earning his degree in 1960. During his medical studies, he also took courses in biology. After a brief tenure at the Children's Hospital of Philadelphia doing virology research, he returned to Germany in 1968 as the head of the newly opened Institute for Virology at the University of Würzburg.

While continuing research on the Epstein-Barr Virus, he shifted his focus toward the study of HPV to uncover its role as an agent causing cervical cancer. In the biopsies of cervical cancer patients, he found a novel HPV-DNA and identified the new tumorigenic HPV16 type in 1983. In 1984, he cloned HPV16 and 18 from patients with cervical cancer. Pathologists were finding HPV types 16 and 18 in about 70% of cervical cancer biopsies.

The global public health burden attributable to HPV is considerable. It is the most common sexually transmitted virus, which is a precursor for cervical cancer. It is found in 99.7% of women with histologically confirmed cervical cancer, affecting about 500,000 women per year.

Luc Montagnier was born in Chabris, in central France, and was educated at the University of Poitiers in Paris. Two years after earning a science degree in 1953, he began his science career as a research assistant and joined the Pasteur Institute in Paris in 1972. Meanwhile, he obtained his medical degree in 1960. After serving as an endowed chair at Queens College in New York for three years, he returned to France in 2001 as Professor emeritus at the Pasteur Institute.

In the early 1980s, Montagnier, working at the Pasteur Institute with a team that included Barré-Sinoussi, identified a retrovirus that was eventually named HIV and shown to be the agent causing AIDS.

There was considerable controversy about whether it was Dr. Robert Gallo from the National Institutes of Health in the United States or Montagnier from the Pasteur Institute in Paris who discovered the AIDS-causing virus. In 1987, the U.S. and the French governments resolved the controversy by agreeing to share the credit for the discovery of HIV. However, the Nobel Foundation did not include Dr. Gallo while awarding the 2008 Nobel Prize in Physiology or Medicine.

Françoise Barré-Sinoussi was born in Paris, France. from a young age, Barré-Sinoussi was interested in analyzing insects and animals' behavior. Under a false assumption that studying medicine would take longer time and more money, she enrolled at the university to study science. Since she had to work to earn money while studying, she attended courses sporadically—yet, with dedication and motivation, she excelled in her studies and obtained her PhD in 1974. She came to the United States as a fellow at the National Institutes of Health briefly before joining Montagnier's unit at the Pasteur Institute. Her collaboration with Montagnier, Jean-Claude Chermann, and other scientists at the Pasteur Institute led her to discover the retrovirus from a lymph node biopsy of an individual who had been suffering from a disease that later came to be known as AIDS.

The discovery of HIV as the etiological agent of AIDS was a pivotal step for developing antiretroviral remedies, which helped control the HIV epidemic and saved tens of millions of lives around the world.

2009

☆
Physics
Charles Kuen Kao, Willard S. Boyle, and George E. Smith

*for groundbreaking achievements concerning the transmission of light
in fibers for optical communication* (to Kao) and *for the invention of an
imaging semiconductor circuit - the CCD sensor* (Boyle and Smith).

☆
Chemistry
Venkatraman Ramakrishnan, Thomas A. Steitz and Ada E. Yonath

for studies of the structure and function of the ribosome.

☆
Physiology or Medicine
Elizabeth H. Blackburn, Carol W. Greider and Jack W. Szostak

*for the discovery of how chromosomes are protected
by telomeres and the enzyme telomerase.*

☆
Literature
Herta Müller

*who, with the concentration of poetry and the frankness
of prose, depicts the landscape of the dispossessed.*

☆
Peace

*Barack H. Obama for his extraordinary efforts to strengthen
international diplomacy and cooperation between peoples.*

☆
Economics
Elinor Ostrom, and Oliver E. Williamson

for her analysis of economic governance, especially the commons
(to Ostrom) and *for his analysis of economic governance,
especially the boundaries of the firm* (to Williamson).

2009

ELIZABETH HELEN BLACKBURN (BORN 1948), CAROLYN WIDNEY GREIDER (BORN 1960) AND JACK WILLIAM SZOSTAK (BORN 1952)

Elizabeth Helen
Blackburn

Carolyn Widney
Greider

Jack William Szostak

Elizabeth H. Blackburn, **Carol W. Greider**, and **Jack W. Szostak** were awarded the Nobel Prize in Physiology or Medicine for their discovery of how chromosomes are protected by telomeres and the enzyme telomerase.

The long, thread-like DNA molecules that carry our genes are packed into chromosomes, the caps on their ends being telomeres. Elizabeth Blackburn and Jack Szostak discovered that a unique DNA sequence in the telomeres protects the chromosomes from degradation. Carol Greider and Elizabeth Blackburn identified "telomerase," the enzyme that makes telomere DNA. These discoveries explained how the telomeres protect the ends of the chromosomes and that they are built by telomerase.

If telomeres are shortened, cells age quickly. If telomerase activity is high, telomere length is maintained, and cells age slowly and live longer. Since cancer cells do not die quickly, the elucidation of how telomeres protect cells and how the telomerase enzyme enables that function became fundamental steps in understanding their longevity—knowledge that was crucial for developing new lines of treatments.

Elizabeth Blackburn was born in Hobart, Tasmania, Australia. After undergraduate studies at the University of Melbourne, she received her PhD in 1975 from the University of Cambridge, England, and was a postdoctoral researcher at Yale University, USA. She was on the faculty at the University of California, Berkeley, and has been a professor of biology and physiology at the University of California, San Francisco, since 1990.

Carol Greider was born in San Diego, California. She studied at the University of California in Santa Barbara and Berkeley, where she obtained her PhD in 1987 with Blackburn as her supervisor. After doing postdoctoral research at Cold Spring Harbor Laboratory, she was appointed a professor in the Department of Molecular Biology and Genetics at Johns Hopkins University School of Medicine in Baltimore in 1997.

Jack Szostak was born in London, UK, and grew up in Canada. He studied at McGill University in Montreal and Cornell University in Ithaca, New York, where he received his PhD in 1977. He has been at Harvard Medical School since 1979, and at the time of his Nobel Prize acceptance, he was a professor of genetics at Massachusetts General Hospital in Boston.

Elizabeth Blackburn learned to map DNA sequences early in her research career. When studying the chromosomes of *Tetrahymena*, a unicellular ciliate organism, she identified a DNA sequence repeated several times at the ends of the chromosomes. The function of the specific adapter sequence named CCCCAA was not known at that time. By then, Szostak had observed that a linear DNA molecule, a type of Mini chromosome, rapidly degrades when introduced into yeast cells.

Blackburn presented her results at a conference in 1980. They caught Szostak's interest, and he and Blackburn decided to perform an experiment to cross the boundaries between distant species. Szostak coupled the CCCCAA sequence to the Mini chromosomes and returned them to yeast cells. The results published in 1982 were striking— the telomere DNA sequence protected the Mini chromosomes from degradation. Later, it became evident that telomere DNA, with its characteristic sequence, is present in most plants and animals, from amoeba to man.

Carol Grieder, then a graduate student, and her supervisor Blackburn, started to investigate if the formation of telomere DNA could be due to an unknown enzyme. On Christmas Day, 1984, Grieder discovered signs of enzymatic activity in a cell extract. Grieder and Blackburn named the enzyme telomerase, purified it, and showed that it consists of RNA and protein. The RNA component turned out to contain the CCCCAA sequence mentioned above.

Some inherited diseases are now known to be caused by telomerase defects, including certain forms of congenital aplastic anemia, in which insufficient cell divisions in the stem cells of the bone marrow cause severe anemia. Telomerase defects also cause certain inherited diseases of the skin and the lungs.

Blackburn, Greider, and Szostak's discoveries helped scientists understand how cells age and die, crucial steps in uncovering the mysteries of disease mechanisms and advancing the development of treatments for previously uncurable conditions.

Many scientists initially wondered if telomere shortening could explain the entire organism's aging process, prompting research to prevent aging. However, it became clear that the aging process is affected by several factors, and the telomere is one of them. Research activities in this field remain intense.

2010

☆

Physics
Andre Geim and Konstantin Novoselov
for groundbreaking experiments regarding the two-dimensional material graphene.

☆

Chemistry
Richard F. Heck, Ei-ichi Negishi, and Akira Suzuki
for palladium-catalyzed cross couplings in organic synthesis.

☆

Physiology or Medicine
Robert G. Edwards
for the development of in-vitro fertilization.

☆

Literature
Mario Vargas Llosa
for his cartography of structures of power and his trenchant images of the individuals resistance, revolt, and defeat.

☆

Peace
Liu Xiaobo for his long and non-violent struggle for fundamental human rights in China.

☆

Economics
Peter A. Diamond, Dale T. Mortensen and Christopher A. Pissarides
for their analysis of markets with search frictions.

2010

SIR ROBERT GEOFFREY EDWARDS (1925-2013)

Robert G. Edwards was awarded the Nobel Prize in Physiology or Medicine for the development of human in-vitro fertilization (IVF). His contributions helped treat infertility, benefiting tens of thousands of couples worldwide.

Robert Edwards was born in Batley, England. After military service in the Second World War, he studied biology at the University of Wales in Bangor and Edinburgh University in Scotland, where he received his PhD in 1955. He joined the National Institute for Medical Research in London in 1958 and initiated his research on human fertilization. In 1963, Edwards moved to Cambridge, where he began his long-term collaboration with Dr. Patrick Christopher Steptoe, a senior obstetrician and colleague who conducted much-needed clinical work in developing IVF.

Edwards' initial research focus in the 1950s was to understand the principles and processes of human fertilization step by step in a systematic approach. Other scientists had shown that egg cells from rabbits could be fertilized in test tubes when sperm was added, giving rise to offspring. Edwards decided to investigate if similar methods could be used to fertilize human egg cells and, if successful, to treat infertility.

But he soon realized that, after all, humans are not like rabbits! Their respective reproductive life cycles differed widely. After a series of well-planned studies in collaboration with Steptoe, Edwards discovered the nature of the maturation of human eggs and the role of maternal hormones in the maturation of human embryos. He also clarified the time point at which eggs become receptive to being fertilized by sperm. Studies by him and others discovered critical experimental conditions that were conducive to activating the sperm so that it could successfully fertilize an egg.

Although Edwards succeeded in fertilizing a human egg in 1969, his efforts to facilitate the fertilized egg's maturation into an embryo failed. That is when Edwards understood that only eggs that mature in the ovary can successfully accept sperm and get fertilized. Thus, he continued his experiments to find the best methods to obtain eggs in the woman's reproductive cycle at the most appropriate time.

And then there were funding obstacles. The United Kingdom's Medical Research Council decided not to continue to fund his research project. However, he sought and obtained private donations from which he managed to continue his work. The topic of research became a hotbed for ethical and philosophical debate. Several important religious leaders, ethicists, and even scientists called for an end to such research work. At the same time, about an equal number of individuals from different disciplines supported this path-breaking research. Despite hurdles, by the mid to late 1970s, Edwards had largely succeeded in achieving this quest of finding the ideal time and method to obtain an egg from a woman's ovary.

In 1977, Lesley and John Brown, a couple who had failed to have a child for over nine years, came to Edward's clinic for consultation. Edwards offered them the IVF. He accomplished fertilization in a culture dish. At the eight-cell embryonic stage, he implanted the embryo in Mrs. Brown's uterus. On July 25th, 1978, doctors delivered Mrs. Brown a healthy baby girl via a Caesarian delivery. The baby girl was full-term and completely healthy. The happy parents named their precious girl Louise, whose birthday has now become a red-letter day in the annals of medicine.

Edward's success in carrying out IVF rekindled the hopes of millions across the world. IVF technology spread quickly. Edwards and Steptoe established the Bourn Hall Clinic in Cambridge, the world's first center for IVF therapy. Obstetricians and gynecologists from around the world came to Bourn Hall to learn IVF methods. Over the years, additional advances were made through research and clinical trials, such that an entirely new branch of obstetrics and gynecology called Reproductive Medicine has flourished since then.

Within a decade, over 2000 children had been born following IVF. IVF technology is now almost half a century old. It has undergone numerous advances, and most reassuringly, children born after IVF therapy have been as healthy as other children.

In 1999, Great Britain issued a 63-penny stamp titled Sculpture of Test-Tube Baby. This stamp was one of the four Millennium Series-3 stamps released as The Patient's Tale. The three other British inventions and discoveries honored in the series include Jenner's vaccination against smallpox, nursing care showing a patient on a trolly, and Alexander Fleming's famous penicillin mold, which led to the development of penicillin.

2011

☆

Physics
Saul Perlmutter, Brian P. Schmidt and Adam G. Riess
for the discovery of the accelerating expansion of the Universe through observations of distant supernovae.

☆

Chemistry
Dan Shechtman
for the discovery of quasicrystals.

☆

Physiology or Medicine
Bruce A. Beutler, Jules A. Hoffmann, and Ralph M. Steinman
for their discoveries concerning the activation of innate immunity (to Beutler and Hoffmann) and *for his discovery of the dendritic cell and its role in adaptive immunity* (to Steinman).

☆

Literature
Tomas Tranströmer
because, through his condensed, translucent images, he gives us fresh access to reality.

☆

Peace
Ellen Johnson Sirleaf, Leymah Gbowee and Tawakkol Karman
for their non-violent struggle for the safety of women and for women's rights to full participation in peace-building work.

☆

Economics
Thomas J. Sargent and Christopher A. Sims
for their empirical research on cause and effect in the macroeconomy.

2011

Bruce Alan Beutler (born 1957), Jules Alphonse Hoffmann (born 1941), and Ralph Marvin Steinman (1943-2011)

Bruce Alan Beutler

Jules Alphonse Nicolas Hoffmann

Ralph Marvin Steinman

The Nobel Prize in Physiology or Medicine was awarded to Bruce A. Beutler and Jules A. Hoffmann for their discoveries concerning innate immunity activation and Ralph M. Steinman for his discovery of the dendritic cell and its role in adaptive immunity. Together, these discoveries revolutionized the knowledge concerning the immune system and the principles for its activation.

When microorganisms such as viruses, bacteria, fungi, or parasites attack man or other animals, their body responds by mounting an immune response. However, the identity of such gatekeepers mounting immune responses remained elusive for a long time. Bruce Beutler and Jules Hoffmann discovered receptor proteins that recognize microorganisms and activate innate immunity, the first step in the body's immune response. This first line of defense can destroy invading microorganisms and trigger inflammation that blocks the assault by the invading agent.

In 1973, Ralph Steinman discovered a new cell type called the dendritic cell. He speculated that dendritic cells could activate T cells—a cell type that has a key role in adaptive immunity and immunologic memory. He showed that dendritic cells initiated vivid responses by T cells in cell culture experiments. If such responses occur when a foreign agent invades the organism, the immune system could react against the invader, avoiding an attack on the body's endogenous molecules. These findings were initially met with skepticism, but subsequent work by Steinman demonstrated that dendritic cells have a unique capacity to activate T cells.

Bruce Beutler was born in Chicago. He received his MD from the University of Chicago in 1981 and worked as a scientist at Rockefeller University in New York. He also held appointments at UT Southwestern Medical Center in Dallas and the Scripps Research Institute in La Jolla, California. At the time of the Nobel Prize, he had rejoined the University of Texas Southwestern Medical Center in Dallas as a professor in the Center for the Genetics of Host Defense.

Jules Hoffmann was born in Echternach, Luxembourg. He studied at the University of Strasbourg in France, where he obtained his PhD in 1969. After postdoctoral training at the University of Marburg, Germany, he returned to Strasbourg, where he headed a research laboratory from 1974 to 2009. He has also served as director of the Institute for Molecular Cell Biology in Strasbourg and, from 2007 to 2008, as President of the French National Academy of Sciences.

Ralph Steinman was born in Montreal, Canada. His family moved to Sherbrooke, Quebec, where his father owned a small clothing store. After high school in that city, Ralph joined McGill University, where he studied science and received his medical degree 1968 from Harvard Medical School. While at Harvard, he spent a year as a research fellow in the laboratory of Elizabeth Hay, who introduced him to cell biology and the immune system. During his stay there, he met Claudia Hoeffel, a medical social worker; they married in 1971.

After completing his medical training, he joined Rockefeller

University in 1970 as a postdoctoral fellow in the Laboratory of Cellular Physiology and Immunology, headed by Zanvil A. Cohn and James G. Hirsch. Steinman's early research, in collaboration with Cohn, attempted to understand how the white cells of the immune system function. Steinman and Cohn discovered dendritic cells, a previously unknown class of immune cells that constantly formed and retracted their processes. This discovery changed the field of immunology.

Thus, the discoveries of the three Nobel Laureates together helped to understand how the innate and adaptive phases of the immune responses are activated. Likewise, a failure to elicit such defense mechanisms also proved insights into the mechanisms of disorders of failed immunity. Such knowledge also helped develop new avenues for the prevention and therapy of infections, cancer, and inflammatory diseases.

On October 3, 2011, the Nobel Committee announced that Steinman was one of the three recipients of that year's Nobel Prize in Physiology or Medicine. But the committee did not know that Steinman had died three days earlier, on September 30, of pancreatic cancer. The statutes of the Nobel Foundation did not allow the prizes to be awarded posthumously. After deliberation, the Nobel Foundation made an unprecedented decision to award the prize since the intention to do so "was made in good faith." His wife, Claudia Steinman, accepted the award on his behalf in December of that year.

Steinman's daughter said that her father had joked the previous week about the possibility of his receiving the Nobel Prize. He had said: "I know I have got to hold out for that. They don't give it to you if you have passed away."

The family donated the money from the Nobel Prize award to charity: $500,000 to the Cohn-Steinman Professorship at Rockefeller University and $250,000 to the Steinman Family Foundation, which supports the careers of young scientists and science education.

2012

☆

Physics
Serge Haroche and David J. Wineland
for ground-breaking experimental methods that enable measuring and manipulation of individual quantum systems.

☆

Chemistry
Robert J. Lefkowitz and Brian K. Kobilka
for studies of G-protein-coupled receptors.

☆

Physiology or Medicine
Sir John B. Gurdon and Shinya Yamanaka.
for the discovery that mature cells can be reprogrammed to become pluripotent.

☆

Literature
Mo Yan
who with hallucinatory realism merges folk tales, history and the contemporary.

☆

Peace
European Union (EU)
for over six decades contributed to the advancement of peace and reconciliation, democracy and human rights in Europe.

☆

Economics
Alvin E. Roth and Lloyd S. Shapley
for the theory of stable allocations and the practice of market design.

2012

SIR JOHN BERTRAND GURDON (BORN 1933) AND SHINYA YAMANAKA (BORN 1962)

Sir John Bertrand Gurdon   **Shinya Yamanaka**

The Nobel Prize in Physiology or Medicine was awarded jointly to Sir John B. Gurdon and Shinya Yamanaka for discovering that mature cells can be reprogrammed to become pluripotent. Their discovery inaugurated new fields of research exploring the mysteries of immune biology.

Humans develop from fertilized egg cells. Immature embryo cells (also called pluripotent stem cells) have the capacity to mature into different cell types meant to carry out specific functions depending upon what tissues they become part of. Such differentiation results in the formation of other cell and tissue types, such as nerve cells, muscle cells, liver cells, and all other cell types, each destined to perform specific functions in the body.

Gurdon challenged the long-held dogma that the life cycle of pluripotent cells is unidirectional—that is, once the cells begin to mature, they cannot reverse their life cycle and cannot become immature cells again. He hypothesized that the pluripotent cell's fate is not destined to a single line of maturation—that they can indeed reverse their life course.

In 1962, he tested this hypothesis by replacing the cell nucleus of a frog's egg cell with a nucleus from a mature, specialized cell derived from the intestine of a tadpole. The egg developed into a fully functional, cloned tadpole, and subsequent experiment repeats yielded adult frogs.

Gurdon's landmark discovery stimulated widespread interest in confirming his findings and promoting the possibility of cloning experimental mammals. While his research helped to accept the possibility that the nucleus of a mature specialized cell can be induced to reverse its life course and become an immature pluripotent cell, it was still not clear if this process could be accomplished without having to remove cell nuclei and implant them into other cell types.

More than 40 years after Gurdon's discovery, Shinya Yamanaka's research filled that knowledge gap. He studied pluripotent stem cells taken from the embryo (also called embryonal stem cells), cultured them in the laboratory, and tried to find the genes that kept them immature. After identifying several of these genes, he tested whether they could reprogram mature cells to become pluripotent stem cells.

Yamanaka and coworkers did this work by introducing these genes in different combinations into the mature cells from connective tissue and fibroblasts and examined the results under the microscope. This approach showed that a combination of genes worked with a surprisingly simple recipe for success. They reprogrammed their fibroblasts into immature stem cells by introducing four genes together. The resulting induced pluripotent stem cells (iPS cells) could develop into mature cells such as fibroblasts, nerve cells, and gut cells.

John Gurdon was born in Dippenhall, a village in Surrey, United Kingdom. He received his Doctorate from the University of Oxford in 1960 and was a postdoctoral fellow at California Institute of Technology. He joined Cambridge University, UK, in 1972 and served as Professor of

Cell Biology and Master of Magdalene College. At the time of the Nobel Prize, Gurdon had been serving at the Gurdon Institute in Cambridge.

Shinya Yamanaka was born in Osaka, Japan. He obtained his MD in 1987 at Kobe University and trained as an orthopedic surgeon before switching to basic research. Yamanaka received his PhD at Osaka City University in 1993, after which he worked at the Gladstone Institutes in San Francisco and Nara Institute of Science and Technology in Japan. At the time of the Nobel Prize, he was a senior investigator at the Gladstone Institutes.

In 2006, Yamanaka and colleagues published their discovery that intact mature cells could be reprogrammed into pluripotent stem cells. Scientists welcomed this discovery as a breakthrough. Reversing the biological life cycle of specialized cells and turning them back to the early pluripotent stage offered new tools for researchers in medical fields. For instance, one can obtain skin cells from patients with skin diseases, reprogram and examine them in the laboratory to determine how they differ from the cells of healthy individuals. Such an insight into the disease mechanism helps in developing new treatment modalities.

2013

☆

Physics
François Englert and Peter W. Higgs
for the theoretical discovery of a mechanism that contributes to our understanding of the origin of mass of subatomic particles.

☆

Chemistry
Martin Karplus, Michael Levitt and Arieh Warshel
for the development of multiscale models for complex chemical systems.

☆

Physiology or Medicine
James E. Rothman, Randy W. Schekman and Thomas C. Südhof
for their discoveries of machinery regulating vesicle traffic, a major transport system in our cells.

Literature
Alice Munro
master of the contemporary short story.

☆

Peace
Organization for the Prohibition of Chemical Weapons (OPCW)
for its extensive efforts to eliminate chemical weapons.

☆

Economics
Eugene F. Fama, Lars Peter Hansen and Robert J. Shiller
for their empirical analysis of asset prices.

2013

James Edward Rothman (born 1950), Randy Wayne Schekman (born 1948) and Thomas Christian Südhof (born 1955)

| James Edward Rothman | Randy Wayne Schekman | Thomas Christian Südhof |

The Nobel Prize in Physiology or Medicine was awarded jointly to **James E. Rothman**, **Randy W. Schekman**, and **Thomas C. Südhof** for their discoveries of machinery regulating vesicle traffic, a major transport system in our cells.

Consider a cell as a biochemical factory producing molecules and transporting them within and outside the cell boundary. The interior of this factory is highly organized. Like a post office that organizes incoming and outgoing mail in appropriate slots, the cellular molecules produced are transferred to appropriate destinations for further modification before being delivered to the cell border for exporting.

For example, the thyroid gland goes through several steps while manufacturing thyroid hormones, which are then released into the bloodstream to exert their actions on tissues in the body. Likewise, the islet cells in the pancreas manufacture insulin and release it into the bloodstream, which controls the concentration of blood sugar. Similarly, signaling molecules called neurotransmitters are sent from one nerve cell to another and transported in miniature bubble-like packages called vesicles, which are released at precise locations for their chemical actions.

The discoveries of the 2013 Nobel Laureates helped uncover the intricate steps involved in the transportation of cellular molecules inside and outside of the cells.

James Rothman was born in Haverhill, a city in Massachusetts, USA. After receiving his PhD from Harvard Medical School in 1976, he worked as a postdoctoral fellow at the Massachusetts Institute of Technology and at Stanford University in California in 1978. At Stanford, he initiated his research on cellular vesicles. He also worked at Princeton University, Memorial Sloan-Kettering Cancer Institute, and Columbia University. At the time of the Nobel Prize, he was the Chairman of the Department of Cell Biology at Yale University New Haven, Connecticut, USA.

Randy Schekman was born in St Paul, Minnesota, USA, and studied at the University of California in Los Angeles and at Stanford University, where he obtained his PhD in 1974 under the supervision of 1959 Nobel Prize winner Arthur Kornberg. At the time of the Nobel Prize award, he was a Professor in the Department of Molecular and Cell Biology at the University of California, Berkeley, which he had joined in 1976.

Thomas Südhof was born in Göttingen, Germany. He studied at the Georg-August-Universität in Göttingen, where he received an MD in 1982 and a Doctorate in neurochemistry the same year. In 1983, he moved to the University of Texas Southwestern Medical Center in Dallas, Texas, as a postdoctoral fellow with Michael Brown and Joseph Goldstein, both of whom shared the 1985 Nobel Prize in Physiology or Medicine. Südhof became an investigator at Howard Hughes Medical Institute in 1991 and was appointed Professor of Molecular and Cellular Physiology at Stanford University in 2008.

Randy Schekman wanted to understand how the intracellular vesicles know where and when to deliver their cargo. In the 1970s, he started using yeast as a model system to study the genetic basis of molecular transport system. He discovered three classes of genes that controlled different stages of cell's transport system, which turned out to be a very tightly regulated system.

While studying vesicle transport in mammalian cells in the 1980s and 1990s, Rothman was also intrigued by the nature of the cell's transport system. He discovered that a protein complex enables vesicles to dock and fuse with their target membranes only at specific locations on the protein. This ensures that the molecular target fuses precisely to the right location and binds to each other as the two sides of a zipper. Such mechanisms ensure that the outer membrane of the cell helps to release the protein-bound molecules into the bloodstream or into an exocrine duct.

Südhof tried to understand the mechanism of nerve cell communication in the brain. The neurotransmitters were known as signaling molecules. They are released from vesicles that fuse with the outer membrane of nerve cells by machinery like those discovered by Rothman and Schekman. But what is the basis of their precise release mechanism?

To understand this critical question, Südhof studied the role of calcium ions and identified calcium sensitive proteins in nerve cells. He discovered the molecular machinery that responds to an influx of calcium ions and directs proteins to bind vesicles to the outer membrane of the nerve cell. The zipper opens and substances carrying signals are released. Südhof's discovery explained how precisely the contents of the vesicles are released when needed.

The discoveries of the 2013 Nobel Laureates explained the processes of intracellular transportation systems. Now scientists believe that defective vesicle transport systems may be the underlying factors for many neurological and immunological disorders, and for diabetes.

2014

☆

Physics
Isamu Akasaki, Hiroshi Amano and Shuji Nakamura
for the invention of efficient blue light-emitting diodes which has enabled bright and energy-saving white light sources.

☆

Chemistry
Eric Betzig, Stefan W. Hell and William E. Moerner
for the development of super-resolved fluorescence microscopy.

☆

Physiology or Medicine
John O'Keefe, May-Britt Moser and Edvard I. Moser
for their discoveries of cells that constitute a positioning system in the brain.

☆

Literature
Patrick Modiano
for the art of memory with which he has evoked the most ungraspable human destinies and uncovered the life-world of the occupation.

☆

Peace
Kailash Satyarthi and Malala Yousafzai
for their struggle against the suppression of children and young people and for the right of all children to education.

☆

Economics
Alfred Jean Tirole
for his analysis of market power and regulation.

2014

JOHN O'KEEFE (BORN 1939), MAY-BRITT MOSER (BORN 1963) AND EDVARD INGJALD MOSER (BORN 1962)

John O'Keefe   May-Britt Moser   Edvard Ingjald Moser

John O'Keefe, **May-Britt Moser**, and **I. Moser** were awarded the Nobel Prize in Physiology or Medicine for their discoveries of cells that constitute a positioning system in the brain.

An awareness of our position in relation to other things around us is an essential feature of our brain health. Only by being aware of where we are at any given time can we plan on our next step of action. As for instance, you need to know that currently you are working in the office, and when the office hours are over, you get ready to leave for home. If you are not aware of your current position in time and space, you would be at a loss.

The sense of place and the ability to navigate are fundamental to our existence. The sense of place gives a perception of position in the environment. During navigation, it is interlinked with a sense of distance that is based on motion and knowledge of previous positions. The need for such an awareness for healthy brain function had long been recognized. But the neurological processes involved in providing such an awareness had been a mystery for a long time.

John O'Keefe was born in New York City. He received his doctoral degree in physiological psychology from McGill University, Canada in 1967. After that, he moved to England for postdoctoral training at University College London. At the time of the Nobel Prize, John O'Keefe had been the Director of the Sainsbury Wellcome Centre in Neural Circuits and Behavior at University College London.

May-Britt Moser was born in Fosnavåg, Norway. She studied psychology at the University of Oslo together with her future husband and co-laureate Edvard Moser. She received her PhD in neurophysiology in 1995. She was a postdoctoral fellow at the University of Edinburgh and subsequently a visiting scientist at University College London before moving to the Norwegian University of Science and Technology in Trondheim in 1996. May-Britt Moser was appointed Professor of Neuroscience in 2000, and at the time of receiving the Nobel Prize, she was the Director of the Centre for Neural Computation in Trondheim.

Edvard Moser was born in Ålesund, Norway. He obtained his PhD in neurophysiology from the University of Oslo in 1995. He was a postdoctoral fellow together with his wife and co-Laureate May-Britt Moser, first at the University of Edinburgh and later as a visiting scientist in John O'Keefe's laboratory in London. In 1996, the couple moved to the Norwegian University of Science and Technology in Trondheim, where Edvard Moser became Professor in 1998. At the time of the Nobel Prize, he had been the Director of the Kavli Institute for Systems Neuroscience in Trondheim.

John O'Keefe was fascinated by the problem of how the brain controls behavior and decided to answer this question using neurophysiological methods. In the late 1960s, he designed experiments to record signals from individual nerve cells in a part of the brain called the hippocampus in rats moving freely in a room. These studies showed him that certain nerve cells were activated when the animal assumed a particular place in the environment, which he called the "place cells." They registered visual input and built up an inner map of the environment. O'Keefe concluded that the hippocampus generates numerous maps, represented by the collective activity of place cells that are activated in different environments. Therefore, the memory of an environment can be stored as a specific combination of place cell activities in the hippocampus.

More than three decades later, in 2005, May-Britt and Edvard Moser discovered another key component of the brain's positioning system. They were mapping the connections to the hippocampus in rats moving in a room when they discovered an astonishing pattern of activity in a nearby part of the brain called the entorhinal cortex. Here, certain cells were activated when the rat passed multiple locations arranged in a hexagonal grid. Each of these cells was activated in a unique spatial pattern, and collectively, these "grid cells" constitute a coordinate system that allows for spatial navigation.

Recent investigations with brain imaging techniques, as well as studies of patients undergoing neurosurgery, have provided evidence that place and grid cells also exist in humans. In patients with Alzheimer's disease, the hippocampus and entorhinal cortex are frequently affected at an early stage, and these individuals often lose their way and cannot recognize the environment. Knowledge about the brain's positioning system may, therefore, help us understand the mechanism underpinning the devastating spatial memory loss that affects people with this disease.

The discovery of the brain's positioning system represents a paradigm shift in our understanding of how groups of specialized cells execute higher cognitive functions. It has opened new avenues for understanding other cognitive processes, such as memory, thinking, and planning.

2015

Physics
Takaaki Kajita and Arthur B. McDonald
for the discovery of neutrino oscillations, which shows that neutrinos have mass

☆

Chemistry
Tomas Lindahl, Paul Modrich and Aziz Sancar
for mechanistic studies of DNA repair

☆

Physiology or Medicine
William C. Campbell, Satoshi Ōmura, and Tu YouYou
for their discoveries concerning a novel therapy against infections caused by roundworm parasites (to Campbell and Omura), and *for her discoveries concerning a novel therapy against Malaria* (to Tu YouYou)

☆

Literature
Svetlana Alexievich
for her polyphonic writings, a monument to suffering and courage in our time

☆

Peace
National Dialogue Quartet
for its decisive contribution to the building of a pluralistic democracy in Tunisia in the wake of the Jasmine Revolution of 2011

☆

Economics
Angus Deaton
for his analysis of consumption, poverty, and welfare

2015

WILLIAM CECIL CAMPBELL (BORN 1930), SATOSHI ŌMURA (BORN 1935), AND TU YOUYOU (BORN 1930)

William Cecil Campbell   **Satoshi Ōmura**   **Tu YouYou**

The Nobel Prize in Physiology or Medicine was awarded to William C. Campbell and Satoshi Ōmura for their discoveries concerning a novel therapy for infections caused by roundworm parasites and to Tu YouYou for her discoveries concerning a novel therapy for malaria.

Parasitic worms, also called helminths, have been known to cause diseases in humans for millennia. Nearly one-third of the world's population has been infested by parasites at some time during their lifespan. The infestations are most common in sub-Saharan Africa, South Asia, and Central and South America. Onchocerciasis (river blindness) causes chronic inflammation in the cornea. Lymphatic filariasis, or elephantiasis, affects more than 100 million people, causing disabling lymphatic swelling of the extremities and scrotum. Besides the stigma, the disease causes severe disability. River blindness and lymphatic filariasis are both caused by parasites.

Malaria is also a parasitic infestation known to mankind since the time of recorded history. The single-cell malaria parasite is transmitted when an infected mosquito bites humans. The parasite invades the body, producing severe fever with chills. It invades and breaks red blood cells, causing chronic anemia. When the malaria parasite invades the brain, the condition is called cerebral malaria, which has a high mortality rate. It causes lasting brain damage among the survivors. Up to 3.4 billion of the persons in the world contract malaria—most victims are young children.

William Campbell was born in Ramelteon, in county Donegal, Ireland. After receiving a BA from Trinity College, University of Dublin in 1952, he received a PhD in 1957 from the University of Wisconsin, Madison, in the US. He then worked at Merck Institute for Therapeutic Research as Senior Scientist and Director for Assay Research and Development. At the time of the Nobel Prize, he had been working as a Research Fellow Emeritus at Drew University in New Jersey.

Satoshi Ōmura was born in the Yamanashi Prefecture, Japan. He received a PhD in Pharmaceutical Sciences in 1968 from the University of Tokyo, Japan, and a PhD in Chemistry in 1970 from Tokyo University of Science. He was a researcher at the Kitasato Institute, where he became Professor in 1975. At the time of the Nobel Prize, he had become Professor Emeritus at Kitasato University.

Tu YouYou was born in China and graduated from the Pharmacy Department at Beijing Medical University in 1955. Beginning as Assistant Professor in 1965 at the China Academy of Traditional Chinese Medicine she rose to the position of Professor in 1985 at the same Institute.

Satoshi Ōmura initially focused his research on *Streptomyces* bacteria that live in soil. Decades earlier, Selman Wakeman had extracted "streptomycin," an anti-tuberculosis antibiotic, from similar soil samples. Wakeman was awarded the 1952 Nobel Prize in Physiology or Medicine for this discovery.

Ōmura isolated new strains of *Streptomyces* from soil samples and successfully cultured them in the laboratory. From many thousand different cultures, he selected about 50 of the most promising, with the intent that they would be further analyzed for their activity against harmful microorganisms and try to isolate new bioactive compounds from this bacterial growth. Out of the 50, one culture turned out to be Streptomyces avermitilis, which became the source of the Avermectin—a precursor for many antiparasitic compounds.

Campbell, an expert in parasite biology, acquired Ōmura's *Streptomyces* cultures, explored their efficacy, and showed that a component from one of the cultures was remarkably efficient against parasites in domestic and farm animals. The initial, purified bioactive agent was named Avermectin. This became the parent compound, the chemical modifications of which resulted in the development of a more effective compound, Ivermectin. This drug was able to kill the parasitic larvae in humans—a great achievement against diseases that had no effective treatment until then.

Tu turned her attention to traditional herbal medicine to tackle the challenge of developing novel malaria therapies. From a large-scale screen of herbal remedies in malaria-infected animals, an extract from the plant *Artemisia annua* emerged as an interesting candidate. Since the benefits of the drug were inconsistent, she revisited the ancient literature and discovered clues that guided her to extract the active component from *Artemisia annua*, which was later called Artemisinin, which was highly effective against the Malaria parasite.

The discoveries of Avermectin and Artemisinin have revolutionized therapy for patients suffering from devastating parasitic diseases. Campbell, Ōmura, and Tu have transformed the treatment of parasitic diseases. The global impact of their discoveries and the resulting benefit to mankind are immeasurable.

In a 2011 interview, Tu YouYou said that her father chose her given name YouYou, adapting it from a Chinese saying that translates as, "Deer bleat *you you* while eating wild Hao (the Artemisia plant)." She added, "How this [her first name] links my whole life with *qinghao* [Artemisia] will probably remain an interesting coincidence forever."

2016

☆

Physics
David J. Thouless, F. Duncan M. Haldane, and J. Michael Kosterlitz

*for theoretical discoveries of topological phase
transitions and topological phases of matter*

☆

Chemistry
Jean-Pierre Sauvage, Sir J. Fraser Stoddart and Bernard L. Feringa

for the design and synthesis of molecular machines

☆

Physiology or Medicine
Yoshinori Ohsumi

for his discoveries of mechanisms for autophagy

☆

Literature
Bob Dylan

*for having created new poetic expressions within
the great American song tradition*

☆

Peace
Juan Manuel Santos

*for his resolute efforts to bring the country's more
than 50-year-long civil war to an end*

☆

Economics
Oliver Hart and Bengt Holmström

for their contributions to contract theory

2016

YOSHINORI OHSUMI (BORN 1945)

Yoshinori Ohsum

Yoshinori Ohsumi was awarded the Nobel Prize in Physiology or Medicine for his discoveries of mechanisms for autophagy, a fundamental process for degrading and recycling cellular components.

The word *autophagy* means self-eating, which originates from the Greek words *auto*, meaning "self," and *phagein*, meaning "to eat." This concept emerged during the 1960s when researchers first observed that the cell could destroy its own contents by enclosing them within membranes, forming sack-like vesicles that are transported to a recycling compartment called the *lysosome* for degradation.

In 1974, the Belgian scientist Christian de Duve was awarded the Nobel Prize in Physiology or Medicine for discovering the lysosomes, a group of enzymes located in the organelles within cells, which help digest proteins, carbohydrates, and lipids. Research in the 1960s revealed that lysosomes contain large amounts of cellular contents, sometimes even whole organelles, indicating lysosomes as the sites for degradation of cells' cargo. Studies also showed a new type of vesicles that transport cellular cargo to lysosomes for degradation. Christian de Duve coined autophagy or "self-eating." The new vesicles were named *autophagosomes*.

Ohsumi was born in Fukuoka, Japan, about six months before the end of World War II. He says that at that time, "…everyone was equally poor." Despite economic hardship during childhood, he kept his dream of becoming a scientist alive. He caught fish and collected and studied insects and plants. Because he was a weak child, his talents in sports were limited. But he would read science books his elder brother would bring from college.

During high school years, Ohsumi became a member of the chemistry club. But over time, he lost interest in chemistry. At the University of Tokyo, he majored in molecular biology. He worked at the molecular biology lab of Kazutomo Imahori. There, under the direction of Akio Maeda, he studied the role of ribosome subunits in protein synthesis. Following the advice of Imahori, Ohsumi pursued cell biology research at G. M. Edelman's lab at Rockefeller University in New York.

Ohsumi's early project was to uncover the mechanism of initiation of DNA replication in yeast cells. His initial attempts to isolate undamaged nuclei failed. But one day, he noticed a white layer at the top of the tube after centrifugation. Instead of discarding it, he looked at that layer under a microscope. He realized that he had isolated highly purified vacuoles—a serendipitous discovery that changed the course of his scientific career.

In 1977, Ohsumi returned to Japan to work at the laboratory of Professor Yasuhiro Anraku at Tokyo University, where he pursued research on the vacuole membrane. This cellular organelle was to inaugurate his career in autophagy research.

Upon starting his lab in 1988, Ohsumi turned his efforts again to studying protein degradation in the vacuoles of yeast cells, which corresponds to the lysosome in human cells. However, initially, he could not definitively identify autophagy under the microscope.

He then considered interrupting the degradation process in the middle, when autophagy, if present, would be active. He discovered that autophagosomes accumulated within the vacuole visible under the microscope, proving autophagy's existence in yeast cells. More importantly, he developed a method to identify and characterize 15 key genes involved in this process. Additional research showed that autophagy is a controlled process involving a cascade of proteins and protein complexes, each regulating a distinct autophagosome initiation and formation stage. Other researchers also showed that autophagy was present in different organisms, including humans, and that the degraded cellular components can be recycled.

Thus, autophagy is both a destructive and constructive activity of the cell that helps in its renewal. For instance, after an infection, autophagy eliminates invading intracellular bacteria and viruses—it is a quality control feature that counteracts the negative consequences of aging.

Ohsumi's discoveries led to a new paradigm in understanding how cells recycle their contents. Mutations in autophagy genes have been associated with cancers and neurological conditions such as Alzheimer's disease, Parkinson's disease, Huntington's disease, and amyotrophic lateral sclerosis.

2017

Physics
Rainer Weiss, Barry C. Barish, and Kip S. Thorne
*for decisive contributions to the LIGO detector and
the observation of gravitational waves*

☆

Chemistry
Jacques Dubochet, Joachim Frank and Richard Henderson
*for developing cryo-electron microscopy for the high-resolution
structure determination of biomolecules in solution*

☆

Physiology or Medicine
Jeffrey C. Hall, Michael Rosbash and Michael W. Young
for their discoveries of molecular mechanisms controlling the circadian rhythm

☆

Literature
Kazuo Ishiguro
*who, in novels of great emotional force, has uncovered the abyss
beneath our illusory sense of connection with the world*

☆

Peace
International Campaign to Abolish nuclear weapons (ICAN)
*for its work to draw attention to the catastrophic humanitarian
consequences of any use of nuclear weapons and for its ground-breaking
efforts to achieve a treaty-based prohibition of such weapons*

☆

Economics
Richard H. Thaler
for his contributions to behavioral economics

2017

JEFFREY CONNOR HALL (BORN 1945), MICHAEL MORRIS ROSBASH (BORN 1944) AND MICHAEL WARREN YOUNG (BORN 1949)

| Jeffrey Connor Hall | Michael Morris Rosbash | Michael Warren Young |

The Nobel Prize in Physiology or Medicine was awarded to **Jeffrey C. Hall, Michael Rosbash and Michael W. Young** for their discoveries of molecular mechanisms controlling the circadian rhythm.

All living organisms have an internal biological clock. That is how our body signals us when to eat, go to bed, and wake up. These events take place in a rhythmic pattern called "circadian rhythm," the phrase with its origin from the Latin words *circa* meaning "around" and *dies* meaning "day." Despite knowing that such a biological clock operates in living organisms, and is in the brain, the mechanisms of its operation were not known.

Our biological clock works with exquisite precision, controlling critical functions, including hormonal cycles, moods and behavior, body temperature, and metabolism. A mismatch between the external environment and the internal clock is the cause of "jet lag," a common condition air travelers experience. However, chronic misalignment can lead to more serious conditions such as chronic irregularity in sleep patterns, emotional instability, impaired decision-making, and reduced cognitive function. The 2017 Nobel Laureates discovered how most members of the animal kingdom adapt their biological clocks and synchronize them with the revolutions of the earth.

Jeffrey Hall was born in New York. He received his doctoral degree in 1971 at the University of Washington in Seattle and did his postdoctoral fellowship at the California Institute of Technology in Pasadena from 1971 to 1973. Then, he joined the faculty at Brandeis University in Waltham, Massachusetts, in 1974. In 2002, he became associated with the University of Maine.

Michael Rosbash was born in Kansas City. He received his doctoral degree in 1970 at the Massachusetts Institute of Technology in Cambridge. During the following three years, he was a postdoctoral fellow at the University of Edinburgh in Scotland. At the time of the Nobel Prize, he was a member of the faculty at Brandeis University.

Michael Young was born in Miami. After his doctoral degree at the University of Texas in Austin in 1975, he did a postdoctoral fellowship at Stanford University. At the time of the Nobel Prize, he was a member of the faculty at Rockefeller University in New York.

In 1729, the French astronomer Jean Jacques d'Ortous de Mairan made the first scientific observation of circadian rhythm by placing a mimosa plant in a light-tight dark room and observed that the plant continued to unfold its leaves in the morning and closed them in the evening. It took 200 additional years for scientific progress in understanding the nature of the biological rhythm when in the 1930s, a German biologist named Erwin Bünning showed that the bean plant leaf movement had a period of 24.4 hours--not 24 hours--in constant light conditions and that the trait can be inherited. He thereby demonstrated the existence of an internal clock that controlled diurnal rhythms that were synchronized by external stimuli.

During the 1970's, Seymour Benzer and his student Ronald Konopka attempted to identify the genes that controlled the circadian rhythm in fruit flies. Together they demonstrated that mutations in an unknown gene located on the x chromosome of Drosophila melanogaster—fruit fly—disrupted their circadian clock. They named this the *period (per)* gene. However, how the *per* gene influenced circadian rhythm remains a mystery.

Another decade had to pass until progress in recombinant DNA technology helped to advance our knowledge about circadian rhythm by successfully cloning the *period* gene—a feat accomplished by Jeffery Hall and Michael Rosbash at Brandies University and Michael Young from Rockefeller University. Hall and Rosbash discovered *period* gene's protein product PER protein accumulated during the night and was degraded during the day. PER protein levels oscillated over a 24-hour cycle, in synchrony with the circadian rhythm.

Although the regularity of the feedback mechanism explained the oscillatory aspect of the circadian rhythm, it was still unclear what controlled the frequency of oscillation. Michael Young's research provided an answer to the missing link: he and his team identified yet another gene, *doubletime*, encoding the DBT protein that delayed the accumulation of the PER protein. This provided an insight into how rhythmic oscillation of the biological clock more closely matched the 24-hour cycle.

The paradigm-shifting discoveries by the three Nobel Laureates established key mechanistic principles that control the biological clock. Over the years, other scientists have discovered additional molecular components of the clockwork mechanism and explained their stability and function. Today, circadian biology has emerged as a dynamic research field with implications for our health and well-being.

2018

☆

Physics
Arthur Ashkin, Gérard Mourou, and Donna Strickland
for groundbreaking inventions in the field of laser physics and for
the optical tweezers and their application to biological systems
(to Ashkin), and *for their method of generating high-intensity,
ultra-short optical pulses* (to Mourou and Strickland)

☆

Chemistry
Frances H. Arnold, George P. Smith and Sir Gregory P. Winter
for the directed evolution of enzymes (to Arnold), and *for the phage
display of peptides and antibodies* (to Smith and Winter)

☆

Physiology or Medicine
James P. Allison and Tasuku Honjo
*for their discovery of cancer therapy by inhibition
of negative immune regulation*

☆

Literature
Olga Tokarczuk
*for a narrative imagination that with encyclopedic passion
represents the crossing of boundaries as a form of life*

☆

Peace
Denis Mukwege and Nadia Murad
*for their efforts to end the use of sexual violence as
a weapon of war and armed conflict*

☆

Economics
William D. Nordhaus, and Paul M. Romer
*for integrating climate change into long-run macroeconomic
analysis* (to Nordhaus), *for integrating technological innovations
into long-run macroeconomic analysis* (to Romer)

2018

JAMES PATRICK ALLISON (BORN 1948) AND TASUKU HONJO (BORN 1942)

James Patrick Allison   **Tasuku Honjo**

The Nobel Prize in Physiology or Medicine was awarded jointly to **James P. Allison and Tasuku Honjo** for their discovery of cancer therapy by inhibiting negative immune regulation.

As the best-selling author Siddhartha Mukherjee puts it, cancer is the "Emperor of all Melodies." Millions of people die each year of cancer. The 2017 Nobel Laureates in Physiology or Medicine established an entirely new approach to treating cancer—they developed methods to stimulate the inherent ability of our immune system to attack tumor cells.

Our immune system is the inherent "army" that discriminates "self" from "non-self, which then mounts an attack to eliminate the invading agent, be it a bacterium, a virus, or a parasite. The key players in the defense system are the T-cells, a type of white blood cell. Specific regions layered on their outer walls, called receptors, bind to structures recognized as non-self, triggering an immune response. Additional proteins, too, help enhance the functions of T-cells. Similarly, there are "breaks" so that the responses of T-cells are not too aggressive, causing harm to host cells. Thus, there is an intricate balance between accelerators and brakes that favors attacking a foreign agent and, at the same time, prevents autoimmune reactions.

James Allison studied the function of a known protein acting as a brake on the immune system. He realized that if one could "release" the break, one could unleash the immune system to attack cancerous tumors. On the other hand, Tasuku Honjo discovered a protein on immune cells that also acted as a break, but through a different mechanism of action.

Allison was born in Alice, Texas. His eighth-grade math teacher inspired him to pursue a career in science and advised him to secure a position in a summer science program funded by the National Science Foundation at the University of Texas in Austin. After earning a Bachelor of Science degree in 1969 and a PhD in 1973, he worked at the Scripps Clinic and Research Foundation in California (1974-1977) and joined MD Anderson Cancer Center in 1984. From then on until 2012, he served in several prestigious organizations, including Memorial Sloan-Kettering Cancer Center, Weill Cornell Graduate School of Medical Sciences, and Howard Hughes Medical Institute.

Tasuku Honjo was born in Kyoto, Japan. In his Nobel Oration, he describes a dramatic episode. On August 1, 1945, when he was about three-and-a-half years of age, his mother carried him on her back and ran breathlessly out of their home that was burning due an American bomb. and he adds, "…two weeks later, World War II ended."

His father was a surgeon and understood the importance of learning English. Therefore, he hired a Japanese woman from Hawaii to teach Tasuku English. While growing up, Tasuku became interested in studying science upon seeing the rings of Saturn on a clear night using a portable telescope. A few years later, he read the biography of the famous microbiologist Hideyo Noguchi, which strongly influenced him to study biology. After graduation, he enrolled in Kyoto University

Medical School and received a postdoctoral position at the Carnegie Institution in Baltimore. from there, he worked in Dr. Pilip Leder's lab at the National Institutes of Health.

Honjo and his collaborators discovered the "programmed cell death-1 (PD-1)" gene in a career spanning several decades. Two biopharmaceutical companies collaborated to manufacture the first batches of PD-1 cancer drugs, which enabled Honjo to conduct early clinical trials in patients with advanced melanoma and lung or renal cancer. In 2014, the Japanese Pharmaceuticals and Medical Devices Agency approved PD-1 antibodies for melanoma treatment, which was soon followed by the U.S. Food and Drug Administration (FDA).

Allison showed that a product called cytotoxic T-lymphocyte antigen 4 (CTLA4) inhibits T-cell responses. They tried to understand how human leukocyte antigens (HLA) and T-cells identified invaders from self. In 1977, Allison and a colleague, G. N. Callahan, reported that some antigens or other proteins prevented the immune system from attacking cancer cells.

Thus, there was evident logic: by blocking the T-cell inhibitory pathways, one could unleash anti-tumor immune responses. These concepts led to the development of drugs targeting T-cell inhibitory pathways, which were called "immune checkpoint" therapies. One such drug is ipilimumab (Yervoy), which received FDA approval in 2011 for treating metastatic melanomas.

Honjo compares the game of golf to scientific research. "…For every shot, you must think how far you will hit, in which direction… whether to add spin to stop on the green. One must avoid big mistakes." Then he says one of his dreams is to achieve a golf game score "equal to my age before I turn 80." We are unsure if he scored 80 or less in a golf game before he turned 80 in January 2022.

2019

☆

Physics
James Peebles, Michel Mayor, and Didier Queloz
for contributions to our understanding of the evolution
of the universe and Earth's place in the cosmos

☆

Chemistry
John B. Goodenough, M., Stanley Whittingham, and Akira Yoshino
for the development of lithium-ion batteries

☆

Physiology or Medicine
William G. Kaelin Jr, Sir Peter J. Ratcliffe and Gregg L. Semenza
for their discoveries of how cells sense and adapt to oxygen availability

☆

Literature
Peter Handke
for an influential work that with linguistic ingenuity has explored
the periphery and the specificity of human experience

☆

Peace
Abiy Ahmed Ali
for his efforts to achieve peace and international cooperation, and in particular
for his decisive initiative to resolve the border conflict with neighboring Eritrea

☆

Economics
Abhijit Banerjee, Esther Duflo and Michael Kremer
for their experimental approach to alleviating global poverty

2019

William G. Kaelin Jr. (born 1957 Sir Peter John Ratcliffe (born 1954) and Gregg L. Semenza (born 1956)

| William G. Kaelin Jr | Sir Peter John Ratcliffe | Gregg Leonard Semenza |

William G. Kaelin Jr., Sir Peter J. Ratcliffe, and Gregg L. Semenza were awarded the Nobel Prize in Physiology or Medicine for their discoveries of how cells sense and adapt to oxygen availability.

All living animals need oxygen to convert consumed food into energy. Oxygen is the source of survival for all creatures--this has been known for centuries. Still, the exact mechanism of how cells adapt to changing oxygen concentrations remains a mystery. The 2019 Nobel Laureates in Physiology or Medicine discovered molecular machinery that regulates the activity of genes responding to changes in oxygen concentrations.

Otto Warburg received the 1931 Nobel Prize in Physiology or Medicine for discovering how cellular mitochondria use oxygen to convert and produce helpful energy through an enzymatic process. This was the first insight into how living cells utilize oxygen.

Humans possess a sensitive structure called the "carotid body," adjacent to large blood vessels on both sides of the neck. These specialized cells sense the changes in blood oxygen concentrations. The discovery of how the carotid bodies send messages directly to the brain to alter our breathing rate based on our blood oxygen levels earned Corneille Heymans the 1938 Nobel Prize. Other physiological responses to low levels of oxygen (hypoxia) include the production of increased levels of a hormone called erythropoietin (EPO produced by the kidney cells) that stimulates the production of red blood cells in the bone marrow in adults—a process called erythropoiesis.

William Kaelin was born in New York City. He studied chemistry and mathematics at Duke University in Durham, North Carolina, and received his medical degree there in 1982. He then did his residency at Johns Hopkins University. In 2002, he became a professor at Harvard Medical School.

Gregg Semenza was born in New York City. After studying medical genetics at Harvard University, he pursued doctoral studies at the University of Pennsylvania, receiving his doctorate there in 1984. He completed pediatric training at Duke University Medical Center and then joined Johns Hopkins University, where he had worked at the time of the 2019 Nobel Prize award.

Semenza and his team studied the EPO gene and how varying oxygen levels regulate it. They used gene-modified mice and showed how specific DNA segments located next to the EPO gene mediated the response to hypoxia. Sir Peter Ratcliffe also studied the O2-dependent regulation of the EPO gene, and both research groups independently found that the oxygen-sensing mechanisms were present in virtually all tissues and many different cell types.

Semenza identified the cellular components mediating EPO responses to hypoxia. In cultured liver cells, he discovered a protein complex that binds to the identified DNA segment in an oxygen-dependent manner. Semenza called this complex the *hypoxia-inducible factor* (HIF), and his team undertook extensive efforts to purify the HIF. In 1995, they identified the genes encoding HIF, and two different DNA-binding proteins or transcription factors, later named HIF-1α and ARNT. It turns out that when oxygen levels are high, cells contain tiny amounts of HIF-1α. Conversely, when oxygen concentrations are low, HIF-1α amounts increase, enabling activation of the EPO gene stimulating erythropoiesis.

Peter Ratcliffe was born in Lancashire, Great Britain. He studied medicine at the University of Cambridge and completed medical school at St Bartholomew's Hospital in London. After additional studies at the Universities of Oxford and Cambridge, Ratcliffe completed doctorate at Cambridge in 1987. He also worked at the University of Oxford and, since 2016, worked at the Francis Crick Institute in London.

William Kaelin, Jr., a researcher studying cancers, was investigating an inherited syndrome, von Hippel-Lindau's (VHL) disease. This genetic disease leads to a dramatically increased risk of certain cancers in families with inherited VHL mutations. Kaelin showed that the VHL gene encodes a protein that prevents the onset of cancer. He also demonstrated that cancer cells lacking a functional VHL gene express abnormally high levels of hypoxia-regulated genes. However, normal levels were restored when the VHL gene was reintroduced into cancer cells. Ratcliffe and colleagues demonstrated that VHL can physically interact with HIF-1α and is required for its degradation at normal oxygen levels.

The discoveries of the three Nobel Laureates honored in 2019 unraveled one of life's fundamental puzzles, helping to develop new approaches to treating anemia, cancers, and other diseases.

2020

☆

Physics
Roger Penrose, Reinhard Genzel, and Andrea Ghez

for the discovery that black hole formation is a robust prediction of the general theory of relativity (to Penrose) and *for the discovery of a supermassive compact object at the center of our galaxy* (to Genzel and Ghez).

☆

Chemistry
Emmanuelle Charpentier and Jennifer A. Doudna

for the development of a method for genome editing.

☆

Physiology or Medicine
Harvey J. Alter, Michael Houghton, and Charles M. Rice

for the discovery of Hepatitis C virus.

☆

Literature
Louise Glück

for her unmistakable poetic voice that with austere beauty makes individual existence universal.

☆

Peace
World Food Programme (WFP)

for its efforts to combat hunger, for its contribution to bettering conditions for peace in conflict-affected areas and for acting as a driving force in efforts to prevent the use of hunger as a weapon of war and conflict.

☆

Economics
Paul R. Milgrom and Robert B. Wilson

for improvements to auction theory and inventions of new auction formats.

2020

HARVEY JAMES ALTER (BORN 1935), SIR MICHAEL HOUGHTON (BORN 1949) AND CHARLES MOEN RICE (BORN 1952)

Harvey James Alter   **Sir Michael Houghton**   **Charles Moen Rice**

Harvey J. Alter, Michael Houghton, and Charles M. Rice received the Nobel Prize in Physiology or Medicine for discovering the hepatitis C virus.

The term "hepatitis" is derived from the Greek *hēpatos*, the genitive of *hepar* for "liver," and "*-itis*" meaning inflammation. As its name indicates, hepatitis is a group of conditions leading to inflammation of the liver, scarring of liver tissue (cirrhosis), and liver cancer. Each year, millions of people around the world are affected by hepatitis and its complications. Epidemiologists estimate hepatitis-related deaths are more than a million each year. The global health burden of hepatitis is comparable to that of tuberculosis and acquired immune deficiency syndrome (AIDS).

Harvey Alter was born in New York. He received his medical degree at the University of Rochester Medical School and trained in internal medicine at Strong Memorial Hospital. In 1961, he joined the National

Institutes of Health (NIH) as a clinical associate, an appointment that enabled him to skip getting drafted by the United States Army. After the NIH fellowship, he completed the second year of medical residency at the University of Seattle and a hematology fellowship at Georgetown University in Washington, DC, where he spent a few years before rejoining the Department of Transfusion Medicine at the NIH Clinical Center as a senior investigator in 1969.

Michael Houghton was born in the United Kingdom. He received his PhD degree in 1977 from King's College London. He joined G. D. Searle & Company before moving to Chiron Corporation, Emeryville, California, in 1982. He relocated to the University of Alberta in 2010; there, at the time of the Nobel Prize, he was the Chair and Professor of Virology, and the Director of the Li Ka Shing Applied Virology Institute.

Charles M. Rice was born in Sacramento. He received his PhD in 1981 from the California Institute of Technology, where he also trained as a postdoctoral fellow between 1981 and 1985. In 1986, he established a research group at Washington University School of Medicine in St Louis, where he was named full professor in 1995. Since 2001, he has been a Professor at Rockefeller University, New York. During 2001-2018, he became the Scientific and Executive Director of the Center for the Study of Hepatitis C at Rockefeller University, where he continued to serve.

In the 1970s, hepatitis A and hepatitis B were discovered. Hepatitis A is transmitted by polluted water or food and has little long-term impact on the patient. Baruch Blumberg discovered that one form of blood-borne hepatitis was caused by a virus that became known as the hepatitis B virus, an agent previously known as Australia antigen. Blumberg received the 1976 Nobel Prize in Physiology or Medicine for this discovery.

During subsequent years, doctors began encountering patients with the clinical and laboratory features of hepatitis, but were negative for hepatitis A or B. Working at the U.S. National Institutes of Health, Alter and his colleagues studied such patients who had received blood transfusions and developed hepatitis, but were negative for hepatitis A or B. These investigators concluded that a previously unknown contagion also transmitted hepatitis, and the agent was a virus—a "non-A or non-B" virus.

Michael Houghton, working for the pharmaceutical firm Chiron, undertook the arduous work needed to isolate the virus's genetic sequence. Houghton and his coworkers created a collection of DNA fragments from nucleic acids found in the blood of an infected chimpanzee. Most of these fragments came from the genome of the chimpanzee itself. Still, after extensive research studies, they found one positive clone derived from a novel RNA virus belonging to the *Flavivirus* family. In 1989, they published on discovering the hepatitis C virus genome.

Scientists were still not sure whether the virus alone could cause hepatitis or needed additional agents. Charles Rice, a researcher at Washington University in St. Louis, and coworkers noted a previously uncharacterized region at the end of the hepatitis C virus genome that they suspected could be important for virus replication. Through genetic engineering, they created an RNA variant of the hepatitis C virus and injected this into the liver of chimpanzees. The virus appeared in the blood of the animals, which also suffered from chronic hepatitis like humans.

In a delightful biographical note available on the Nobel Foundation Website, Alter describes an ingenious series of epidemiologic sleuthing he undertook while diagnosing the cause of renal failure in a patient of his, who was a truck driver. It seems another patient in the same room showed Alter an article from *Readers Digest* that described renal failure arising from inhaling carbon tetrachloride (CCL4). Alter went to the sleeping segment of the cab where his patient used to sleep and found a leaky fire extinguisher that contained CCL4. The patient had been inhaling the fumes of CCL4, causing his renal failure. Alter says his discovery attained him a heroic status—albeit briefly. Although an intern, his seniors invited him to give the Grand Rounds about his discovery.

The discovery of the hepatitis C virus led to the development of tests to screen blood donors for this virus, virtually eliminating post-transfusion hepatitis in many parts of the world.

2021

☆

Physics
Syukuro Manabe, Klaus Hasselmann, and Giorgio Parisi

for groundbreaking contributions to our
understanding of complex physical systems

☆

Chemistry
Benjamin List and David W.C. MacMillan

for the development of asymmetric organocatalysis

☆

Physiology or Medicine
David Julius and Ardem Patapoutian

for their discoveries of receptors for temperature and touch

☆

Literature
Abdulrazak Gurnah

for his uncompromising and compassionate penetration
of the effects of colonialism and the fate of the refugee
in the gulf between cultures and continents

☆

Peace
Maria Ressa and Dmitry Andreyevich Muratov

for their efforts to safeguard freedom of expression, which
is a precondition for democracy and lasting peace

☆

Economics
David Card, Joshua D. Angrist and Guido W. Imbens

for his empirical contributions to labor economics (to Card),
and *for their methodological contributions to the analysis*
of causal relationships (to Angrist and Imbens).

2021

DAVID JAY JULIUS (BORN 1955) AND ARDEM PATAPOUTIAN (BORN 1967)

David Jay Julius

Ardem Patapoutian

David Julius and Ardem Patapoutian received the Nobel Prize in Physiology or Medicine for their discoveries of receptors for temperature and touch.

How do we sense our environment? How do our eyes detect an object, and our brain recognizes it? How do our inner ears receive sound waves, and our brain deciphers the source, the pitch, and the nature of the sound? How do chemical compounds interact with the receptors in our nose and mouth, resulting in us experiencing the sense of smell and taste?

Joseph Erlanger and Herbert Gasser received the 1944 Nobel Prize in Physiology or Medicine for their discovery of different types of sensory nerve fibers that react to distinct stimuli, such as responses to painful and non-painful touch. Since then, scientists have discovered highly specialized nerve cells that detect and transmit signals of differing types of stimuli, allowing a nuanced perception of our surroundings by our brain.

David Julius was born in New York, the grandson of a couple who fled Czarist Russia and antisemitism in pursuit of a better life in the USA. He grew up in Brighton Beach, a seaside Brooklyn neighborhood which, he says, "was dense and somewhat gritty." He studied at Abraham Lincoln High School, which was the alma mater for writers such as Arthur Miller and Joseph Heller, performers such as Mel Brooks and Beverly Sills, and scientists such as Arthur Kornberg, Paul Berg, and Jerome Karle—the three Nobel Laureates prior to David Julius from that school.

After high school, Julius enrolled at MIT and was drawn towards studying biology and chemistry. He completed graduate studies at Berkeley. In 1989, he accepted a position at the University of California, San Francisco. He says that his "Eureka" moment came when Michael Caterina joined his group and spearheaded efforts to identify the capsaicin receptor, later called TRPV1, using an elegant expression cloning scheme. Together with others, the team was able to identify TRPV1 as a heat-activated ion channel, providing a cogent molecular explanation for a widely appreciated psychophysical experience – the 'hotness' of chili peppers.

Julius gives credit to the National Institutes of Health as a "great engine" driving biomedical research and training in the US.

Ardem Patapoutian is of Armenian descent, born in Beirut, Lebanon. Although Beirut was a beautiful city, because of deteriorating infrastructure and continuous clashes between religious factions many Armenian families began to flee Beirut.

While studying at the American University of Beirut, Patapoutian encountered a frightful experience. One day, a group of militants held him captive, suspecting him of being a spy, and threatened to shoot at his knee. Eventually, they realized that he was harmless and released him.

"But that was the final straw for me," says Patapoutian.

Within a few months, he flew to Los Angeles. He planned on enrolling at the University of California (UCLA). However, he worked for one year delivering pizzas and writing horoscopes for an Armenian newspaper. At UCLA, working at Professor Judy Lengyel's lab, Patapoutian developed a love for research. After majoring in biochemistry and biology, he pursued graduate studies at the California Institute of Technology (Caltech), where he met Nancy Hong, his future wife, a biology undergraduate.

He secured a postdoctoral fellowship at the University of California, San Francisco, where he worked for 5 years before joining Scripps Research in San Diego and Novartis Research Foundation, a nonprofit associated with the Swiss drug company, where Patapoutian established his own lab.

Focusing his research on sensory stimuli, he and his team tried to answer: "How does a temperature-sensitive neuron become *thermossensor*, while neighboring neurons become specialized in sensing touch?" They also tried to understand how physical stimuli, such as temperature and pressure, elicited awareness in the brain.

Following the example of David Julius' demonstration of TRPV1 ion channel activated by heat, Patapoutian tested if other TRP ion channels were involved in thermos-sensation. Within a few years, they identified a channel activated by both cold temperatures and menthol, the cooling compound derived from mint leaves. They called the channel TRPM8. Over the next couple of years, they identified other somatosensory ion channels, including TRPV3 and TRPA1.

After a decade of research on thermos-sensation, Patapoutian and his group, especially his postdoctoral fellow Bertrand Coste, spearheaded a new assay technique to determine which cell types respond to pressure. Coste attempted to record the electrical activity of the cells while poking them with a glass probe. If the cells expressed a mechanically activated channel, they would see changes to the current when they were pushed. Coste found that Neuro2A cells were exquisitely mechanosensitive and decided to focus on these cells. He worked through a list of 300 genes expressed in Neuro2A cells to see if any of them encoded an ion channel. After struggling for one year, Coste discovered that the removal of the 72nd candidate gene wiped out the current—an unstated "Eureka moment." They named the gene PIEZO1 after the Greek word *piezi* for pressure.

Patapoutian was the first Nobel laureate of Armenian origin and the first from Lebanon—thus, both communities celebrated his award.

2022

☆

Physics
Alain Aspect, John F. Clauser, and Anton Zeilinger
for experiments with entangled photons, establishing the violation of Bell inequalities and pioneering quantum information science

☆

Chemistry
Carolyn R. Bertozzi, Morten Meldal and K. Barry Sharpless
for the development of click chemistry and bioorthogonal chemistry

☆

Physiology or Medicine
Svante Pääbo
for his discoveries concerning the genomes of extinct hominins and human evolution

☆

Literature
Annie Ernaux
for the courage and clinical acuity with which she uncovers the roots, estrangements and collective restraints of personal memory

☆

Peace
Ales Bialiatski, The Russian human rights organization Memorial, and the Ukrainian human rights organization Center for Civil Liberties.

☆

Economics
Ben S. Bernanke, Douglas W. Diamond and Philip H. Dybvig
for research on banks and financial crises

2022

SVANTE PÄÄBO (BORN 1955)

Svante Pääbo

The Nobel Prize in Physiology or Medicine was awarded to Svante Pääbo for his discoveries concerning the genomes of extinct hominins and human evolution.

Humanity has always been intrigued by its origins. Where do we come from, and how are we related to those who came before us? What makes us, *Homo sapiens*, different from other hominins? In his 1859 publication, "On the Origin of Species by Natural Selection..." Charles Darwin answered some of these questions, providing geological, archeological, and macro-level biological evidence.

In 2010, Svante Pääbo offered molecular-level evidence connecting modern humans to extinct hominids. He sequenced the genome of the Neanderthal, an extinct relative of today's humans, and in the process, discovered a previously unknown hominin, *Denisova*. He also showed that gene transfer occurred from the extinct hominins to *Homo sapiens* following the migration out of Africa around 70,000 years ago.

Svante Pääbo was born in Stockholm, Sweden. He grew up with his mother, Karin Pääbo, an Estonian chemist who fled her country because of the Soviet invasion in 1944 and arrived in Sweden as a refugee. Svante was born through an extramarital affair with his mother by Sune Bergström (1916–2004), a Swedish biochemist, who also received the Nobel Prize in Physiology or Medicine in 1982.

In 1975, Pääbo started studies at Uppsala University and earned his PhD there. He did postdoctoral research at the Institute of Molecular Biology II at the University of Zurich, Switzerland. In 1987, he continued his postdoctoral fellowship at the University of California, Berkeley, where he began researching the genome of extinct mammals. He accepted the position of professor of general biology at the University of Munich in 1990. Seven years later, Pääbo became the founding director of the Max Planck Institute for Evolutionary Anthropology in Leipzig, Germany.

At the new Institute, Pääbo and his team steadily improved the methods for isolating and analyzing DNA from archaic bone remains. The research team exploited new technical developments, which made DNA sequencing highly efficient. Pääbo also engaged several critical collaborators with expertise in population genetics and advanced sequence analyses. His efforts were successful.

In February 2009, at the Annual Meeting of the American Association of the Advancement of Science (AAAS) in Chicago, Max Planck Institute announced that it had completed the first draft version of the Neanderthal genome in which they had sequenced over 3 billion base pairs in collaboration with the 454 Life Sciences Corporation. Comparative analyses demonstrated that the most recent common ancestor of Neanderthals and *Homo sapiens* lived around 800,000 years ago.

In March 2010, Pääbo and his coworkers published a report about the DNA analysis of a finger bone found in the Denisova Cave in Siberia. These results suggested that the bone belonged to an extinct member of the genus *Homo* that was not yet been recognized, which was named the Denisova hominin. In May 2010, Pääbo and his colleagues also published a draft sequence of the Neanderthal genome. They concluded that there was probably interbreeding between Neanderthals and Eurasian (but not Sub-Saharan African) humans. This concept has general mainstream support. This admixture of modern human and Neanderthal genes is estimated to have occurred roughly between 50,000 and 60,000 years ago in the Middle East.

Pääbo and his coworkers could now investigate the relationship between Neanderthals and modern-day humans from different parts of the world. Comparative analyses showed that DNA sequences from Neanderthals were more like sequences from contemporary humans originating from Europe or Asia than to contemporary humans originating from Africa. This means that Neanderthals and *Homo sapiens* interbred during their millennia of coexistence. In modern-day humans of European or Asian descent, approximately 1-4% of the genome originates from the Neanderthals.

Research provided evidence that the anatomically modern human, *Homo sapiens,* first appeared in Africa approximately 300,000 years ago. At the same time, our closest known relatives, Neanderthals, developed outside Africa and populated Europe and Western Asia from around 400,000 years until 30,000 years ago, at which point they went extinct. About 70,000 years ago, groups of *Homo sapiens* migrated from Africa to the Middle East, and, from there they spread to the rest of the world. *Homo sapiens* and Neanderthals thus coexisted in large parts of Eurasia for tens of thousands of years.

Pääbo's discoveries have generated new understanding of our evolutionary history. At the time when *Homo sapiens* migrated out of Africa, at least two extinct hominin populations inhabited Eurasia. Neanderthals lived in western Eurasia, whereas Denisovans populated the eastern parts of the continent. During the expansion of *Homo sapiens* outside Africa and their migration east, they not only encountered and interbred with Neanderthals but also with Denisovans.

Pääbo 's seminal research led to an entirely new scientific discipline: *paleogenomics*. By revealing genetic differences that distinguish all living humans from extinct hominins, his discoveries provide the basis for exploring what makes us uniquely human. Pääbo 's 2014 book *Neanderthal Man: In Search of Lost Genomes* tells the story of the research effort to map the Neanderthal genome combined with his thoughts on human evolution.

2023

☆

Physics
Pierre Agostini, Ferenc Krausz and Anne L'Huillier
for experimental methods that generate attosecond pulses of light for the study of electron dynamics in matter

☆

Chemistry
Moungi G. Bawendi, Louis E. Brus and Aleksey Yekimov
for the discovery and synthesis of quantum dots

☆

Physiology or Medicine
Katalin Karikó and Drew Weissman
for their discoveries concerning nucleoside base modifications that enabled the development of effective mRNA vaccines against COVID-19

☆

Literature
Jon Fosse
for his innovative plays and prose which give voice to the unsayable

☆

Peace
Narges Mohammadi
for her fight against the oppression of women in Iran and her fight to promote human rights and freedom for all

☆

Economics
Claudia Goldin
for having advanced our understanding of women's labor market outcomes

2023

Katalin Karikó   **Drew Weissman**

The Nobel Prize in Physiology or Medicine was awarded jointly to **Katalin Karikó and Drew Weissman** for their discoveries concerning nucleoside base modifications that enabled the development of effective mRNA vaccines against COVID-19.

The use of vaccines to prevent infections has been available through much of the 20th century. Typically, vaccines using partially or completely inactivated viruses help to stimulate antibody production, helping to prevent infections from those viruses. However, vaccine experts had difficulty in developing vaccines against RNA viruses because RNA-based vaccines elicited intense inflammatory reactions in the host prior to eliciting an antibody response.

In 2005, Katalin Karikó and Drew Weissman discovered that certain modifications of the building blocks of RNA prevented unwanted inflammatory reactions and increased the production of desired antibody proteins. The discovery laid the foundation for effective mRNA vaccines against COVID-19 during the pandemic that began in early 2020.

Karikó co-founded and was CEO of RNARx from 2006 to 2013. From 2013 to 2022, she was associated with BioNTech RNA Pharmaceuticals. In 2022, she left the company to devote more time to research.

Katalin Karikó was born in Szolnok, a city located on the banks of the Tisza River in the heart of the Great Hungarian Plain and grew up in Kisújszállás, a small home without running water, a refrigerator, or television. Karikó excelled in science during her primary education and obtained her BSc degree in biology in 1978 and PhD in biochemistry in 1982 from the University of Szeged. She did her postdoctoral research at the Institute of Biochemistry, Biological Research Center (BRC) of Hungary.

In 1985, BRC lost its funding, forcing Karikó to seek else outside of Hungary. She accepted a research position offered by Robert J. Suhadolnik from Temple University in Philadelphia and moved to the US with her husband and two-year-old daughter. She says that she stuffed £900 into her daughter's teddy bear of they earned by selling their car and exchanging Hungarian currency to British sterling on the black market.

After 3 years working as a postdoctoral fellow at Temple University, she accepted a job at Johns Hopkins University without telling her advisor, Suhadolnik. Upon learning this, Suhadolnik was so upset that he reported to the US immigration authorities that Karikó had been in the US illegally. However, before her extradition order was enforced, she found a position at the Naval Hospital of the Uniformed Services University in Bethesda, Maryland. There, she worked for one year. In 1989, she joined the University of Pennsylvania. However, despite repeatedly trying, she failed to secure research grants, which caused the university not to offer her a tenure-track position. She stayed there until 2013. She joined BioNTech RNA Pharmaceuticals, where she became vice president and later senior vice president.

Since 2021, she has been a Professor at Szeged University and an

Adjunct Professor at Perelman School of Medicine at the University of Pennsylvania. At the time of the Nobel Prize, Karikó had been a professor at the University of Szeged in Hungary.

In 1997, she met Drew Weissman, a professor of immunology who had recently joined the University of Pennsylvania. Once by chance, they met near a photocopier. This meeting initiated a long and fruitful professional collaboration.

A major problem with the therapeutic use of mRNA was that it led to inflammatory reactions. A key insight came about when Karikó focused on why mRNA caused inflammatory reactions, whereas transfer RNA (tRNA), used as a control in experiments, was noninflammatory. Karikó and Weissman discovered that nucleoside modifications in mRNA by replacing uridine with pseudo-uridine reduced immune response. Their manuscript with this key finding was rejected by the prestigious *Nature* and *Science* journals but eventually accepted by the journal *Immunity*.

Another important achievement by the researchers was the development of a delivery technique to package the mRNA in lipid nanoparticles, a novel pharmaceutical drug delivery system for mRNA. The mRNA is injected into tiny fat droplets (lipid nanoparticles), which protect the fragile molecule until it reaches the desired area of the body. They demonstrated its effectiveness in animals.

Drew Weissman was born in Lexington, Massachusetts. He went to Lexington High School, graduating from there in 1977. He obtained his BA and MA degrees from Brandeis University majoring in biochemistry and enzymology. After a residency at Beth Israel Deaconess Medical Center, he did a fellowship at the National Institutes of Health under the supervision of Anthony Fauci, then the Director of the National Institute of Allergy and Infectious Diseases.

Weissman and Karikó overcame another major obstacle by developing a delivery technique to package the mRNA in lipid nanoparticles, a novel pharmaceutical drug delivery system for mRNA that protects the fragile molecule until it can reach the desired area of the body. They demonstrated the effectiveness of the delivery system in animals.

Weissman's laboratory continues to actively research the use of mRNA for next-generation vaccines, gene editing, and mRNA therapeutics. His projects include the development of pan-coronavirus vaccines, gene editing technology to enable genes that produce missing antibodies, and treatments for acute inflammatory conditions. Weissman hopes that mRNA technology can be used to develop vaccines against influenza, herpes, and HIV.

The research studies Karikó and Weissman, were groundbreaking in that they have changed our understanding of how mRNA interacts with our immune system, and enabled development of mRNA vaccines at an unprecedented speed during one of the greatest threats to human health in modern times.

2024

Physics
John J. Hopfield and Geoffrey Hinton

*for foundational discoveries and inventions that enable
machine learning with artificial neural networks*

☆

Chemistry
David Baker, Demis Hassabis, and John Jumper

for computational protein design (to Baker), *and for protein structure
prediction*
(to Hassabis and Jumper)

☆

Physiology or Medicine
Victor Ambros and Gary Ruvkun

*for the discovery of microRNA and its role in post-
transcriptional gene regulation*

☆

Literature
Han Kang

*for her intense poetic prose that confronts historical
traumas and exposes the fragility of human life*

☆

Peace
Nihon Hidankyo

*for its efforts to achieve a world free of nuclear weapons and for demonstrating
through witness testimony that nuclear weapons must never be used again*

☆

Economics
Daron Acemoglu, Simon Henry Roberts
Johnson and James A. Robinson

for studies of how institutions are formed and affect prosperity

2024

VICTOR R. AMBROS (BORN 1953) AND GARY BRUCE RUVKUN (BORN 1952)

Victor R. Ambros   **Gary Bruce Ruvkun**

The Nobel Prize in Physiology or Medicine was awarded jointly to **Victor Ambros and Gary Ruvkun** for the discovery of microRNA and its role in post-transcriptional gene regulation.

Our genome is like an instruction manual for the cells in our body, sending the same instructions to all cells. Yet, the cells produce different products depending upon the cell type. For example, muscle and nerve cells have very distinct characteristics and produce different products. How is this possible despite receiving the same set of original instructions?

In 1993, Victor Ambros and Gary Ruvkun discovered microRNA, a new class of tiny RNA molecules that play a crucial role in gene regulation. These discoveries were later determined to be of fundamental importance in the cellular development and function of organisms.

Ambros was born in New Hampshire and grew up in Vermont. He earned his PhD at MIT under the supervision of David Baltimore, then an MIT professor of biology, who received a Nobel Prize in 1973. Ambros was a longtime faculty member at Dartmouth College before joining the University of Massachusetts Chan Medical School faculty in 2008.

Ruvkun was born in Berkley, California. He received his Bachelor of Arts (BA) degree with a major in biophysics from the University of California, Berkeley in 1973. He earned his PhD in biophysics from Harvard University in 1982. Ruvkun completed postdoctoral research with Robert Horvitz at the Massachusetts Institute of Technology and Walter Gilbert of Harvard—both of whom were Nobel Laureates.

In the 1960s, many scientists discovered specialized proteins known as transcription factors. These proteins can bind to specific regions of DNA and control the flow of genetic information by determining which mRNAs are produced. Since then, thousands of transcription factors have been identified, and for a long time, it was believed that the main principles of gene regulation had been solved.

However, in 1993, Ambrose and Ruvkun published unexpected findings describing a new level of gene regulation that was highly significant and conserved throughout evolution. Their research began in the 1980s with Ambros and Ruvkun, who were postdoctoral fellows in Robert Horvitz's laboratory. There, they studied the genes of the one mm-long roundworm *C. elegans.*

Ambros analyzed the lin-4 mutant in his newly established laboratory at Harvard University. Methodical mapping allowed the cloning of the gene and led to an unexpected finding. The lin-4 gene produced an unusually short RNA molecule that lacked a code for protein production. These surprising results suggested that this small RNA from lin-4 was responsible for inhibiting lin-14.

Concurrently, Ruvkun investigated the regulation of the lin-14 gene in his newly established laboratory at Massachusetts General Hospital and Harvard Medical School. Ruvkun showed that it is not the production of mRNA from lin-14 that is inhibited by lin-4—such inhibition appeared at a later stage in the process of gene expression through the shutdown of protein production. Experiments also revealed a segment in lin-14 mRNA necessary for its inhibition by lin-4.

Ambros and Ruvkun compared their findings, which resulted

in a breakthrough discovery. The short lin-4 sequence matched complementary sequences in the critical segment of the lin-14 mRNA. Ambros and Ruvkun performed further experiments showing that the lin-4 microRNA turns off lin-14 by binding to the complementary sequences in its mRNA, blocking the production of lin-14 protein.

A new principle of gene regulation, mediated by a previously unknown type of RNA, microRNA, had been discovered. They published their findings in two articles in the journal *Cell* in 1993. However, the scientific community did not welcome their findings enthusiastically. Many wondered that while the results were interesting, the unusual mechanism of gene regulation was considered a peculiarity of *C. elegans*, likely irrelevant to humans and other more complex animals.

That perception changed in 2000 when Ruvkun's research group published their discovery of another micro-RNA, encoded by the let-7 gene. Unlike lin-4, the let-7 gene was highly conserved and present throughout the animal kingdom. This article sparked great interest, and hundreds of different microRNAs were identified over the following years.

Abnormal regulation by microRNA can contribute to cancer, and mutations in genes coding for microRNAs have been found in humans, causing conditions such as congenital hearing loss and eye and skeletal disorders. Mutations in one of the proteins in a gene called DICER1 required for microRNA production result in the DICER1 syndrome. This rare genetic disorder increases the risk of both benign and cancerous tumors of the kidney, thyroid, ovary, cervix, testicle, brain, eye, and lining of the lung. The tumors may be benign or cancerous.

Because of the discoveries by Ambros and Ruvkun, scientists around the world began studying microRNAs. Today we know that humans have more than a thousand genes for different micro RNAs, and that microRNAs are present among all multicellular organisms. Thus, the world of microRNA is an endless ocean waiting to be explored to help us understand their role in health and disease.

ABOUT THE AUTHOR

TONSE NARAYANA KRISHNA RAJU, a pediatrician and a neonatologist, served as Professor of Pediatrics at the University of Illinois in Chicago for over three decades. He has written five books of fiction and two translations in Kannada, a classical Dravidian language from Karnataka, South India. His translations include the Kannada translation of the novel Thousand Cranes (Ankita Press, 2010) by the Japanese writer Yasunari Kawabata, the winner of Nobel Prize in Literature; and an English translation of Kannada originals, Poems and a Novella by A. K. Ramanujan (Oxford University Press, 2006).

A member of the American Osler Society, Dr. Raju has written and taught graduate and undergraduate students medical history topics throughout his career. His most recent publications include The Importance of Having a Brain: Tales from the History of Medicine (Expanded second edition, Auctorem 2025) and Don't Stand in Front of a Palace or Behind a Horse: An Illustrated Book of South Indian Proverbs (Auctorem 2025).

www.ingramcontent.com/pod-product-compliance
Lightning Source LLC
Chambersburg PA
CBHW031838200326
41597CB00012B/186